DEVELOPMENTAL AND CELL BIOLOGY SERIES

EDITORS

P. W. BARLOW   P. B. GREEN   C. C. WYLIE

ORGANOGENESIS OF THE KIDNEY

ORGANOGENESIS OF THE KIDNEY

LAURI SAXÉN
Department of Pathology, University of Helsinki

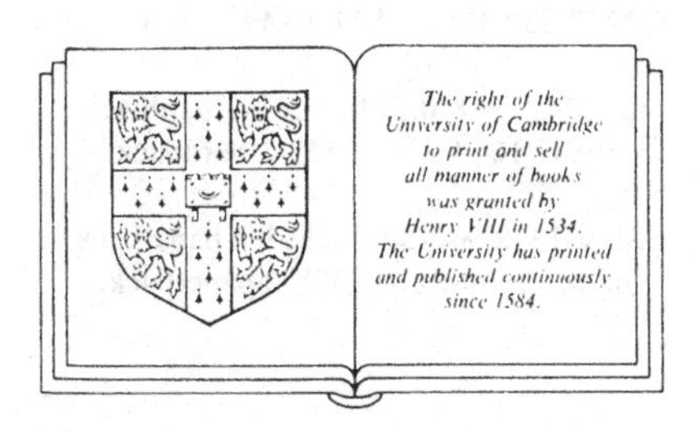

CAMBRIDGE UNIVERSITY PRESS
Cambridge
London New York New Rochelle
Melbourne Sydney

CAMBRIDGE UNIVERSITY PRESS
Cambridge, New York, Melbourne, Madrid, Cape Town, Singapore, São Paulo

Cambridge University Press
The Edinburgh Building, Cambridge CB2 2RU, UK

Published in the United States of America by Cambridge University Press, New York

www.cambridge.org
Information on this title: www.cambridge.org/9780521301527

First published 1987
This digitally printed first paperback version 2006

A catalogue record for this publication is available from the British Library

Library of Congress Cataloguing in Publication data

Saxén, Lauri.
Organogenesis of the kidney.
(Developmental and cell biology series)
Bibliography
Includes index.
1. Kidneys – Growth. I. Title. II. Series.
[DNLM: 1. Kidney – growth & development. WJ 301 S2720]
QP249.S39 1987 599′.0149 86–17590

ISBN-13 978-0-521-30152-7 hardback
ISBN-10 0-521-30152-1 hardback

ISBN-13 978-0-521-03508-8 paperback
ISBN-10 0-521-03508-2 paperback

Contents

Preface

Research in developmental biology, as in many other fields of basic biology, is carried out largely in model-systems. In these, relatively simple events are singled out from the complex situation *in vivo* and examined under controllable conditions. Such model-systems should be simple enough to allow a meaningful analysis, but sufficiently complex to warrant extrapolation of the results to the normal condition *in vivo*. The vertebrate kidney is one such model-system for cell differentiation and organogenesis. It is my hope that the observations and conclusions quoted in this book could be of wider interest to developmental biologists, and that the use of the kidney system could unravel general developmental principles applicable to other cells and organs.

My work on kidney development was initiated during a postdoctoral stay in the laboratory of Clifford Grobstein at Stanford University. The research was continued in Helsinki in close collaboration with my late friend Tapani Vainio. Never, since then, have I worked with a more stimulating scientist, and many of Tapani's ideas guided our research long after his accidental death in 1965.

The 'Wetterkulla Medical Center' (WMC), an informal biannual working symposium, has met regularly during the last 25 years. These occasions have been of great inspiration to me, and they have also provided time for me to write this book. I thank my friends at the WMC: Sulo Toivonen, Osmo Järvi, Kari Penttinen, Erkki Saxén, Harri Nevanlinna, Esko Nikkilä, Kari Cantell and the late Bo Thorell.

I wish to express my deep gratitude to all my students, research fellows and colleagues, whose results and ideas I have quoted in this monograph and whose material I have used extensively for illustration. My special thanks are due to Eero Lehtonen, who has in many ways contributed to the text and the illustrations and who has read the manuscript critically. From the very beginning of the kidney project I have been privileged to work with a most skilful technical team, led and instructed by Anja Tuomi, whom I thank most cordially. I have had the pleasure of collaborating with Arto Nurmi, artist and medical illustrator, for two decades, and I wish to thank him again for his fine artwork. Last but not least, I thank my secretary Annikki Kaitila, whose interest, experience and skill in editing scientific text have been of major importance in finishing this book.

L.S.

Wetterkulla, Eräjärvi
December 1985

1

Ontogenesis of the vertebrate excretory system

Introduction

The functional unit of the vertebrate excretory system, the nephron, is similar in its essential features in all vertebrate classes from cyclostomes to mammals, and it has been found in the fossils of the oldest known vertebrates, the ostracoderms (Torrey, 1965). The nephron consists of: the vascular loop of the glomerulus, the capsule, the nephrocoel, the nephric duct, the nephrostome, and the tubule (Fig. 1.1). These structures may vary in detail, but the differences between the classes are created mainly by the spatial assembly of the nephrons within the organism. Consequently, it has become customary to distinguish between three spatially and temporally different excretory 'organs', the pronephros, the mesonephros, and the metanephros, though many authors consider this classification to be useful merely for descriptive purposes within a continuum of tissue (Fraser, 1950; Fox, 1963; Torrey, 1965). Hence, the term 'holonephros' has occasionally been proposed to cover

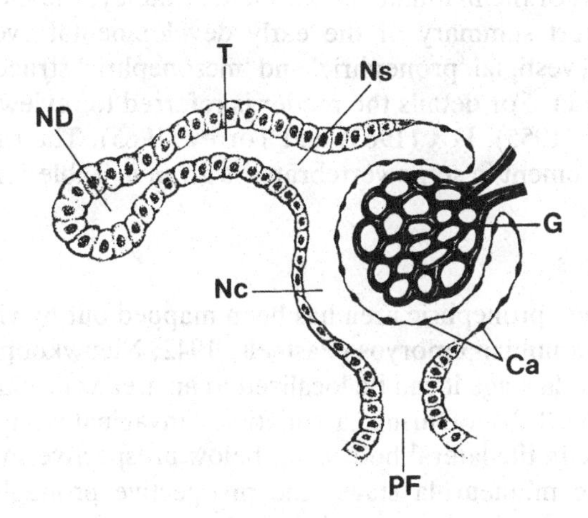

Fig. 1.1. Scheme of nephron (after Torrey, 1954). Ca, capsule; G, glomerulus; Nc, nephrocoel; T, tubule; ND, nephric duct; Ns, nephrostome; PF, peritoneal funnel.

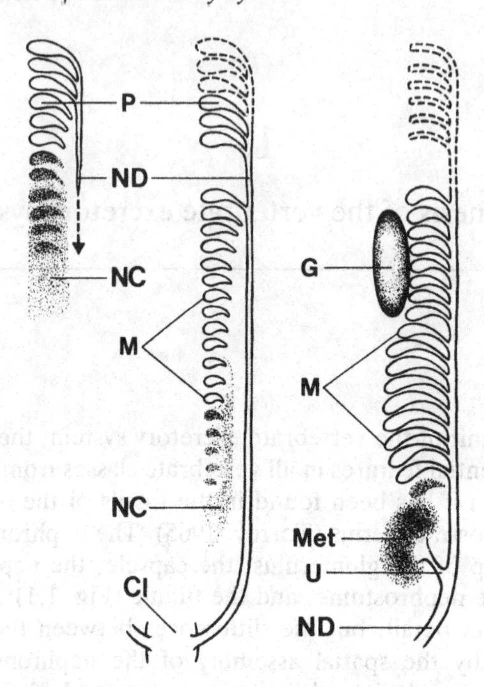

Fig. 1.2. An overall scheme of the development of the vertebrate kidney (after Burns, 1955). P, pronephros; ND, nephric duct; NC, nephrogenic cord; G, gonad; M, mesonephros; Met, metanephros; U, ureter; Cl, cloaca.

the entire temporal and spatial (also phylogenetic) sequence of tubular arrangement within the excretory system (Fig. 1.2). Understanding of the embryogenesis of the mammalian permanent kidney, the metanephros, requires a short summary of the early developmental events of the transitory and vestigial 'pronephric' and 'mesonephric' structures and of the nephric duct. For details the reader is referred to reviews by Fraser (1950), Burns (1955), Fox (1963) and Torrey (1965). The time-table of kidney development in some vertebrates is given in Table 1.1.

The pronephros

The prospective pronephric area has been mapped out by vital staining methods in amphibian embryos (Pasteels, 1942; Nieuwkoop, 1947). At the early gastrula stage it can be localized to an area ventrolateral to the blastopore (Fig. 1.3A). During gastrulation it invaginates and is found in early neurulae in the lateral body wall, below prospective somites 4 and 5. During the midneurula stage, the prospective pronephric area is determined and will form nephric tubules when transplanted to heterotopic sites. Other parts of the embryo have already lost this potency (Fales, 1935).

Table 1.1. *Time-course (in days) of kidney development in some vertebrates*

	Pro-nephros	Meso-nephros	Ureteric bud	Bud with cap	Meta-nephros	Gestational period
Man	22	24	28	32	35–37	267
Macaque	—	—	29–30	31–32	38–39	167
Guinea pig	16	17	20	21	23	67
Rabbit	8.5	9.0	11.5	13	14	32
Rat	10	11.5	12.3	12.5	12.5	22
Mouse	8	9.5	11	—	11	19
Hamster	—	8.5	9.3	9.5	10	16
Chick	1.5	2.3	4	5	6	21

After Hoar & Monie, 1981.

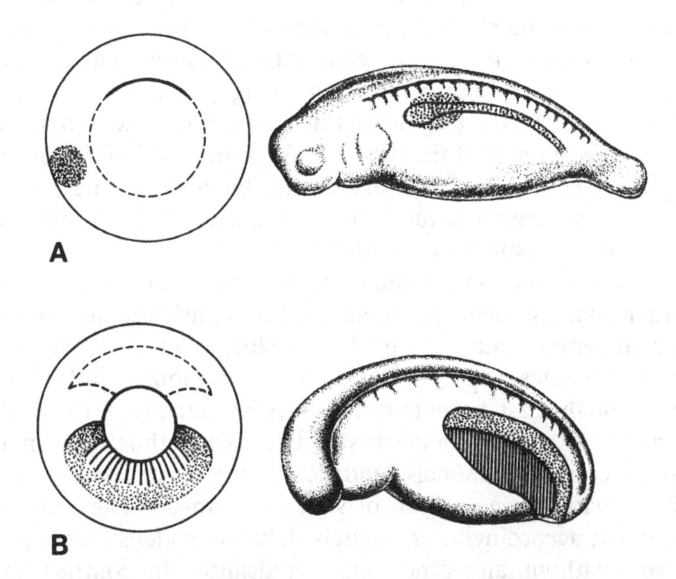

Fig. 1.3. Schemes of the prospective nephric tissue in amphibian gastrulae and the localization of the nephric material in tail-bud-stage embryos. A. The prospective pronephric area in an axolotl gastrula and its subsequent distribution as followed by vital stain (after Pasteels, 1942). B. The mesonephric material (stippled) and the lateral plate tissue (lined) in a salamander gastrula and at the tail-bud stage (after Nieuwkoop, 1947).

The pronephric nephrotomes are located between the somitic mesenchyme and the lateral plate at a level caudal to the third somite. The level and the number of the pronephric tubules vary in different species. The tubules open through the nephrostome into the coelomic cavity, and their free distal ends may fuse to form the most anterior portion of the pronephric duct (p. 6). At the time of regression of the first pronephric tubules, isolated (external) glomeruli develop by budding from the coelomic epithelium and become vascularized by vessels from the dorsal aorta. In the chick embryo, these glomeruli develop on days 3 to 5 at the level of the ninth to the fifteen somites (Davies, 1950). Scanning and transmission electron-microscopic examination of these external glomeruli revealed a histogenesis similar to that in the functional mesonephros (Jacob *et al.*, 1977). The authors therefore conclude that the external glomeruli show the true structural maturation of a functional unit.

The development of the pronephros into a functional organ is restricted to the lower vertebrates and has not been shown conclusively in amniotes. Apart from a few exceptions (some teleosts), the significance of the pronephros as a functional excretory organ is limited to the embryonic (larval) stages, and the structures regress during later development. The functional significance of the pronephros was first suggested by experiments where unilateral pronephrectomy resulted in compensatory hypertrophy of the remaining pronephros (Howland, 1921; Hiller, 1931; Fox, 1956; Bosshard, 1971). It is difficult to estimate the precise time of onset of the function. Rappaport (1955) followed the specific gravity of *Rana pipiens* larvae after bilateral nephrectomy and concluded that an excretory function was required after Shumway stage 18, soon after the larvae became motile.

In all amniotes, the development of the anterior end of the nephrogenic mesenchyme remains rudimentary. Pronephric tubules have been described in reptile embryos and in the chick, but as Torrey (1965) critically emphasizes, 'the transition between tubules judged to be pronephric and the first mesonephric tubules is so gradual as to cast doubt on the reality of the pronephric variety'. The author is even more sceptical concerning mammals, and denies the existence of a proper pronephros, with the exception of some marsupials. The pronephros should consist, accordingly, of 'vaguely defined condensations, vesicles or grooves' without any functional significance. In contrast to this generalization, Toivonen (1945) has described well-formed pronephric tubules in 10-day rabbit embryos at the level of the sixth to the twelfth somites and partially overlapping the mesonephric ducts (Fig. 1.4).

The nephric duct

The nephric (pronephric) duct constitutes the central component of the excretory system throughout development. It is the drainage channel of the functional pronephros and mesonephros, it gives rise to the ureteric bud of the metanephros and ultimately contributes to the male genital system as the ductus deferens. Moreover, as described in Chapter 3, the epithelium of the nephric duct and its derivatives determine the

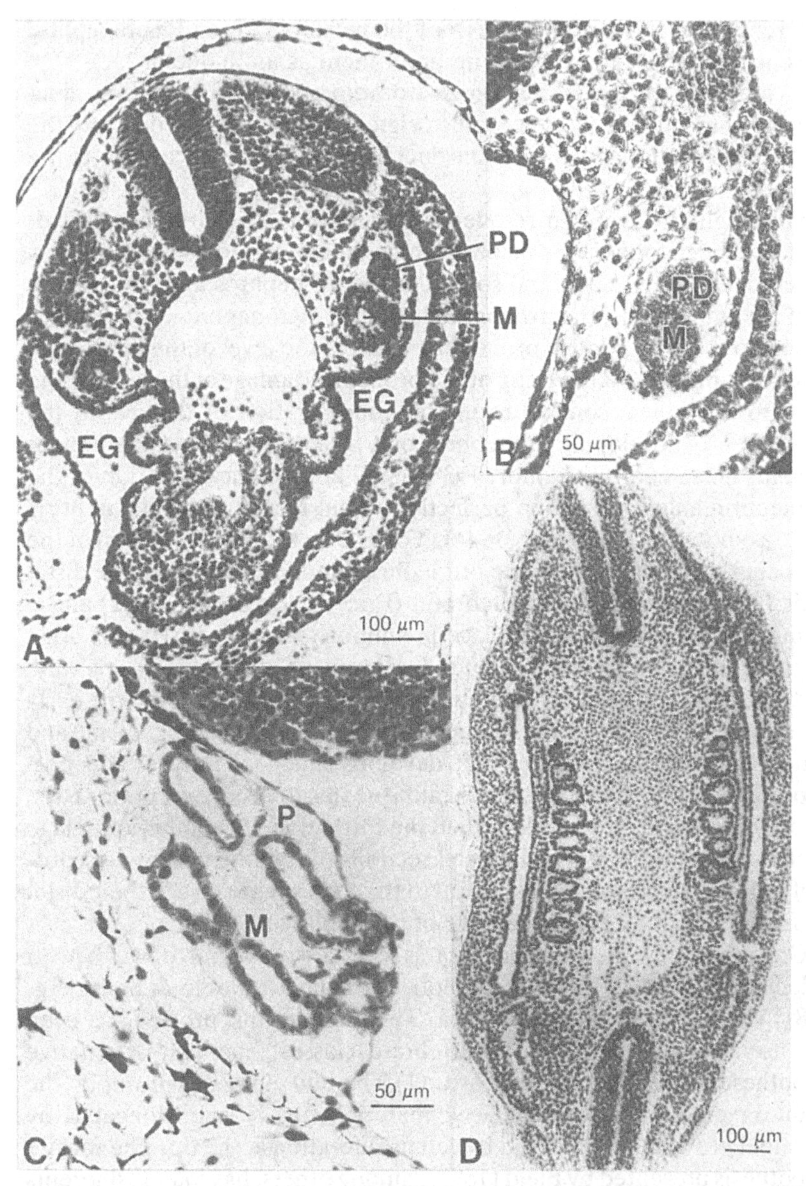

Fig. 1.4. Micrographs of rabbit embryos, demonstrating the early stages of the nephric rudiment (Toivonen, 1945, and unpublished results; courtesy of Dr S. Toivonen). P, pronephros; M, mesonephros; PD, pronephric (nephric) duct; EG, external glomeruli. A. Transverse section of a 10-day rabbit embryo showing the localization of the pronephric duct, the mesonephros and two external glomeruli. B. A detail at the level of segment 20, showing the nephric duct and the mesonephros opening into the abdominal cavity. C. Appearance of a pronephric and a mesonephric nephron at the same level in a rabbit embryo. D. Longitudinal section through a 10-day rabbit embryo showing the nephric duct and the segmental mesonephric nephrons.

differentiation and morphogenesis of the mesonephric and metanephric mesenchymal blastemas by acting upon them as an 'inductor'.

Two aspects of the formation and development of the nephric duct have become objects of interest: the origin of the pronephric duct and the mode of elongation (growth) of the duct from its anterior location into the cloaca.

The original view of an ectodermal origin of the nephric duct (Field, 1891) has been found to be inconsistent with later observations that show conclusively the mesodermal source of the pronephros and the nephric duct. In amphibians, the primary anlage of the pronephric duct is found caudal to the presumptive pronephric area, at the level of the fifth to the seventh somites. Vital staining of the pronephric anlage at the level of the third to the fourth somites results in concentration of the dye in the subsequently developing pronephros only, whereas the duct component remains unstained (O'Connor, 1938). Similarly, surgical deletion of the pronephric anlage results in perfectly normal formation of the nephric duct, as shown by Holtfreter (1944). Yet, the most anterior portion of the duct might still derive from the pronephric tubules by fusion of the distal ends. This occurs in lower vertebrates (Price, 1897; Brauer, 1902) and is frequently suggested to take place in amniotes as well. It remains open whether the avian pronephric duct development is initiated in this way, and it has, in fact, been disputed by Torrey (1965). He explored the early development of the nephric duct in mammals, including man, and concluded that the nephric duct develops independently of the pronephros (Torrey, 1954). If so, the rabbit seems to be an exception to the rule. Toivonen (1945) has described the formation of the anterior anlage of the nephric duct as seen in serial sections of rabbit embryos. Accordingly, the pronephric tubules caudal to the sixth somite fuse at their distal ends and constitute the first anlage of the nephric duct.

Once formed, the nephric duct is progressively laid down with a velocity of 1 mm/20 h (in the axolotl, according to Poole & Steinberg, 1981) until it opens into the cloaca. The mode of this process has been extensively studied in many vertebrate classes, and two alternative hypotheses have been put forward. (1) The bud elongates *in situ* by the addition of new mesodermal cell material, or (2) it is elongated by terminal growth implemented by cell proliferation at the tip. The former hypothesis presented by Field (1891), among others, has found adherents and indirect experimental support (e.g. Goodrich, 1930; van Deth, 1946; Shin-Iké, 1955), but it is not consistent with other experimental data. Such experiments, performed both in amphibian and in avian embryos, have included vital staining methods, deletion operations, and transplantation experiments, followed by light- and electron-microscopic recording of the results. Such investigations have marshalled plenty of evidence for the second hypothesis of the elongation of the nephric duct, the idea of independent growth (Burns, 1938; Gruenwald, 1937, 1942; van Geertruyden, 1942; Holtfreter, 1944; Spofford, 1945; Cambar, 1952*a*, *b*; Calame, 1959, 1962; Croisille *et al.*, 1976). However, if the view of 'independent growth' of the nephric duct is accepted, two alternatives still remain, namely growth by cell multiplication or elongation through cell streaming. Of course, the two are not mutually exclusive.

The importance of an actual proliferation of the terminal epithelial cells of the duct has been disputed by Overton (1959), after a careful analysis of the duct's mitotic rate as a function of stage of development and site. No terminal accumulation of mitoses was found, and the mitotic index, uniform throughout the length of the duct, did not differ significantly from the overall high proliferation rate of the embryos. Hence, the author attributed the elongation of the nephric duct primarily to a migratory capacity of the cells. Her own observations *in vitro* on isolated epithelial fragments of the ducts demonstrated this migration capacity of the duct cells thus confirming similar observations on nephric duct cells transplanted to heterotopic sites (Maschkowzeff, 1936; Nieuwkoop, 1947; Bijtel, 1948, 1968).

Acceptance of the 'migration hypothesis' should lead to exploration of the principles and forces guiding this 'conducted migration' (Weiss, 1947; Trinkaus, 1984). The role of the surrounding tissues in this event had been suggested by Holtfreter (1944), who emphasized the possible significance of the mesoderm, and especially the vascular endothelium, in guiding the migration. This view was shared by Ti-Chow-Tung & Su-Hwei-Ku (1944), who, after rotation and transplantation experiments, concluded that a predetermined pathway was created by the surrounding cells.

Poole & Steinberg (1981) have conducted a thorough scanning electron-microscopic study of the elongation of the pronephric duct in the axolotl (Fig. 1.5). The duct primordium segregates from the dorsal portion of the lateral plate mesoderm at the level of the second to the seventh somites and subsequently extends along the ventrolateral border of the somites. The elongation is mainly due to a rearrangement of cells seen as a thinning of the duct and a reduction of the number of cells across the duct's diameter. Morphology of the duct cells suggests an active migration of those at the tips of the duct. Subsequent, extensive transplantation experiments by the authors (Poole & Steinberg, 1982) have shed more light on the actual guiding forces of this migratory event. Pronephric duct primordia were transplanted into various sites on the flank mesoderm ventral to the primary duct of the host, and their fate was followed by scanning electron microscopy. The transplants showed active migration towards the primary duct (Fig. 1.6), with which they ultimately fused. The distant effect of various tissues previously suggested to act as attractants was eliminated by proper extirpation experiments, and

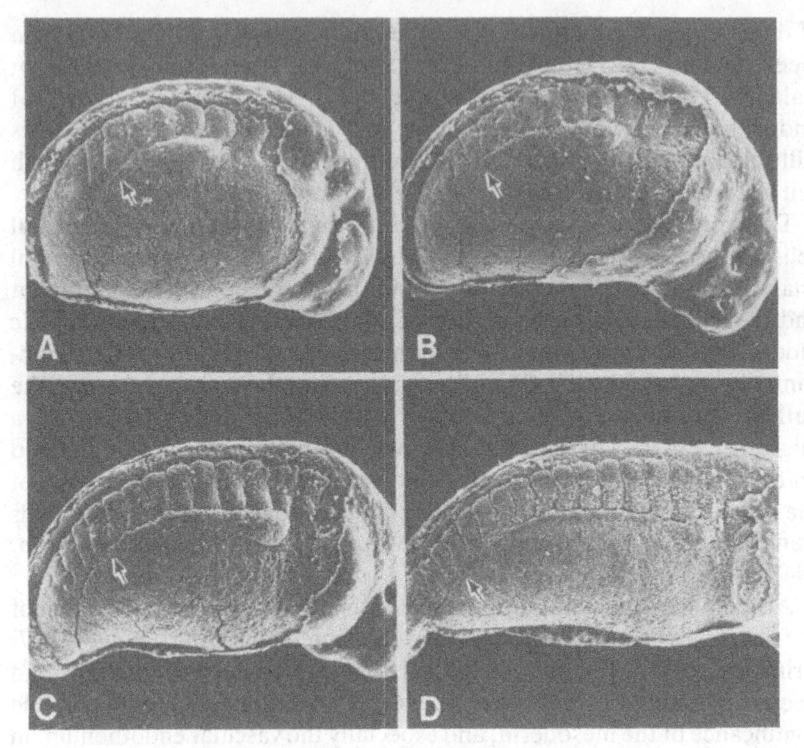

Fig. 1.5. Scanning electron micrographs of the development of the nephric duct in an axolotl embryo. Arrows indicate the tip of the migrating duct (Poole & Steinberg, 1981; courtesy of Dr M. Steinberg).

direction by preformed paths became less plausible as a result of direct observations on the behaviour of the transplanted cells. The authors conclude that the guidance of the migrating duct cells is implemented by a gradient of adhesion provided by the mesoderm, with which the epithelial cells are in contact through thin philopodia. This dynamic gradient is modulated during embryogenesis in relation to the craniocaudal wave of differentiation, and it provides local directional clues for the migrating cells.

The mesonephros

The second segmental excretory organ, the mesonephros, has been thoroughly studied in amphibians and in birds (Burns, 1955; Croisille, 1976) and in some domestic mammals (Tiedemann, 1976, 1979, 1983; Tiedemann & Wettstein, 1980). The prospective mesonephric tissue has been mapped out in amphibian embryos at the late neurula stage (Fig. 1.3B). The tissue is already determined, but it is still flexible, because the

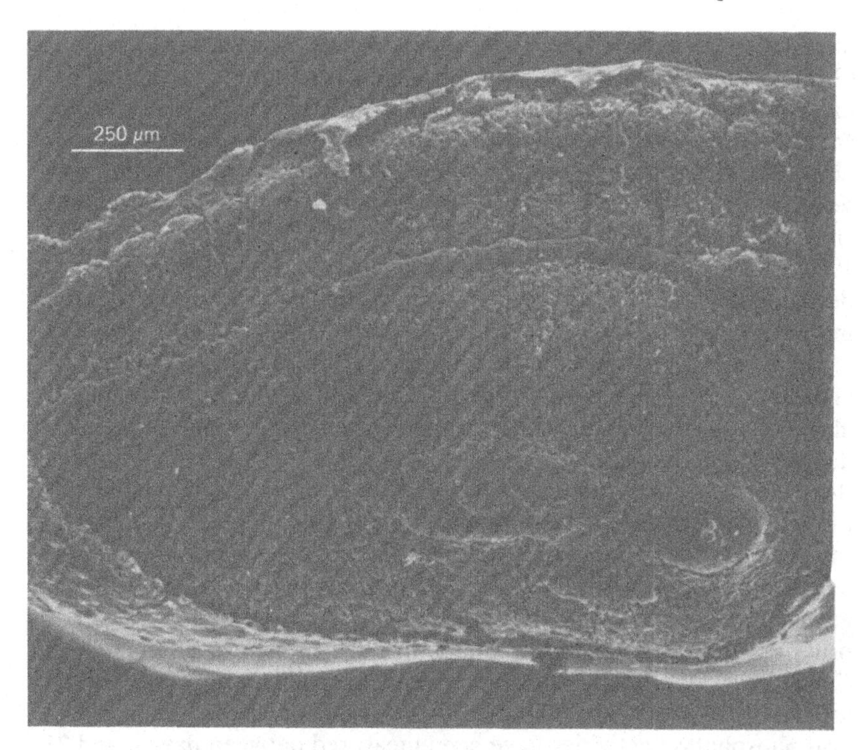

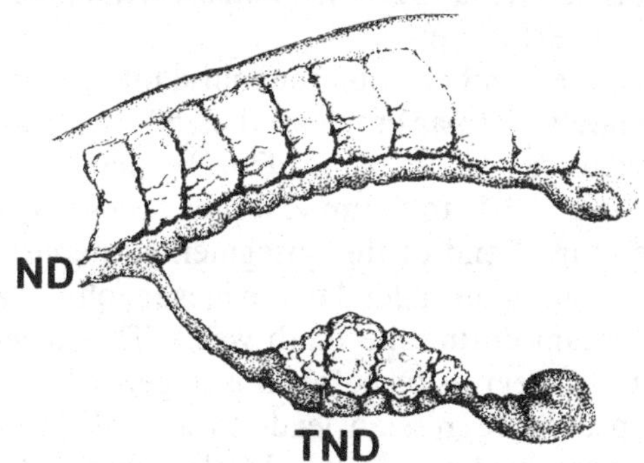

Fig. 1.6. Scheme of a grafting experiment demonstrating the oriented migration of a transplanted nephric duct (TND) and its fusion with primary duct (ND) (Poole & Steinberg, 1982; courtesy of Dr M. Steinberg).

type and arrangement of the tubules can be affected by new surroundings (Machemer, 1929). At the tail-bud stage determination becomes fixed and mesonephric structures will develop in grafts at various sites of the embryo (Humphrey, 1928). This apparently terminal determination occurs only after the pronephric duct has reached the prospective mesonephric mesenchyme, and the sequence suggests an integrated morphogenesis of the two cell lineages. As will be reviewed in Chapter 3 (Table 3.1, p. 53), there is ample evidence favouring this view of an inductive interaction between the nephric cord mesenchyme and the caudally extending pronephric duct. Observations have been reported, however, which strongly suggest that the nephric duct epithelium is a terminal inductor acting upon a predetermined mesonephrogenic mesenchyme with a 'kidney bias' created during its early development (Saxén, 1970*a*). In higher vertebrates, the mesonephros is a transient organ that develops caudally to the pronephros but might also partially overlap the most caudal part of the pronephros (Toivonen, 1945).

There are great differences in size, distribution, and functional maturity of the mesonephric nephrons between the different avian and mammalian species. In the chick, the first (primary) nephrons are seen as depressions in the mesonephric blastema on day 3. During the following three days, new nephrons are sequentially induced and accumulated into the organ in a craniocaudal sequence. Five distinct populations of these tubules have been described that differ in their spatial distribution, length, and stage of maturation (Friebová-Zemanová & Contcharevskaya, 1982). Their total number has been estimated to be of the order of 100 (Bremer, 1915; Stampfli, 1950; Friebová-Zemanová & Concharevskaya, 1982). The regression of the chick mesonephros starts around day 14 (Haffen, 1951), and profound changes in the cathepsin and acid phosphatase activities have been measured between day 14 and 21 (Salzgeber & Weber, 1966).

In mammals, the number of mesonephric tubules and glomeruli vary from approximately 30 (man) to more than 50 (sheep and pig) (Bremer, 1915; for a review, see Tiedemann, 1976). In man, the first mesonephric tubules are found in 3.5- to 4-mm embryos about day 25, and they are located at the cranial end of the unsegmented nephric blastema. New renal vesicles are gradually added to the mesonephros, and their amount reaches a maximum during the ninth week. The most cranial tubules, however, start their regression before this stage, around the fifth week. In female embryos, the regression leads to a total disappearance of the organ during the third month of development, while, in the male embryos, some caudal tubules and the mesonephric duct persist and contribute to the male genitals (efferent ductules and vasa deferentia) (for reviews, see Marin-Padilla, 1964; Du Bois, 1969; Tiedemann, 1976). There is no direct proof of the functional maturity of the human mesonephric nephrons, and an affirmative view is based on merely structural findings (Silverman, 1969).

Morphogenesis of the mesonephric nephron starts with the formation of mesenchymal condensates, which soon develop into renal vesicles and S-shaped bodies – much like similar developmental events in the

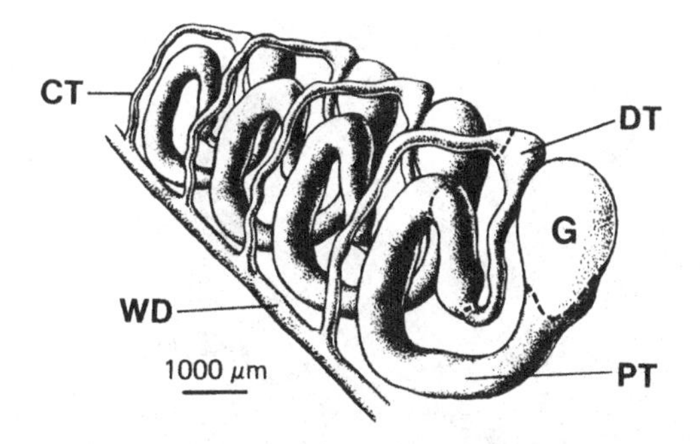

Fig. 1.7. Scheme of the mesonephric nephron (after Schiller & Tiedemann, 1981; Tiedemann, 1976). CT, collecting tubule; WD, Wolffian duct; DT, distal tubule; PT, proximal tubule; G, glomerulus.

metanephros that will be dealt with in more detail below. The nephric body then becomes connected to the pronephric (Wolffian) duct, after which the following segments can be distinguished (Fig. 1.7): the mesonephric corpuscle (prospective glomerulus), the proximal tubule with three portions, the distal tubule and the collecting tubule (duct). These segments show distinct differences detectable at the ultrastructural level or by histochemical and immunochemical methods (Croisille, 1969, 1970, 1976; Croisille *et al.*, 1971, 1974; Martin *et al.*, 1971; Tiedemann, 1976, 1983; Tiedemann & Wettstein, 1980; Schiller & Tiedemann, 1981; Tiedemann & Zaar, 1983). The origin of these different segments of the nephron has been the object of some controversy. Most authors consider the entire nephron to be of mesenchymal origin, but some believe that the collecting duct is derived from the Wolffian duct (for reviews, see Hamilton 1952; Wolff, 1969). To solve the problem, Croisille and his collaborators prepared chick/quail chimaeric kidneys and followed the fate of their cells by using the quail 'nuclear marker' to distinguish between chick and quail cells (Le Douarin & Barq, 1969). Chimaeras were constructed by combining the anterior half of chick blastoderm with the posterior half of a quail embryo at the same stage of development. At this stage, the pronephric (Wolffian) duct was still restricted to the anterior half, while the undifferentiated nephrogenic mesenchyme was found in the posterior half. The chimaeric recombinants were cultivated *in vitro* until the nephric duct had invaded the mesonephrogenic (quail) mesenchyme. Then the chimaeric anlages were subgrafted intracoelomically and analysed histologically. The analysis showed that the cells of the glomeruli, the proximal tubules, and of the distal tubules were exclusively of quail origin, i.e. derived from the mesenchymal blastema. Only 'very few, extremely short' collecting ducts were shown to be derivatives of the chick Wolffian duct (Gumpel-Pinot *et al.*, 1971; Croisille *et al.*, 1974;

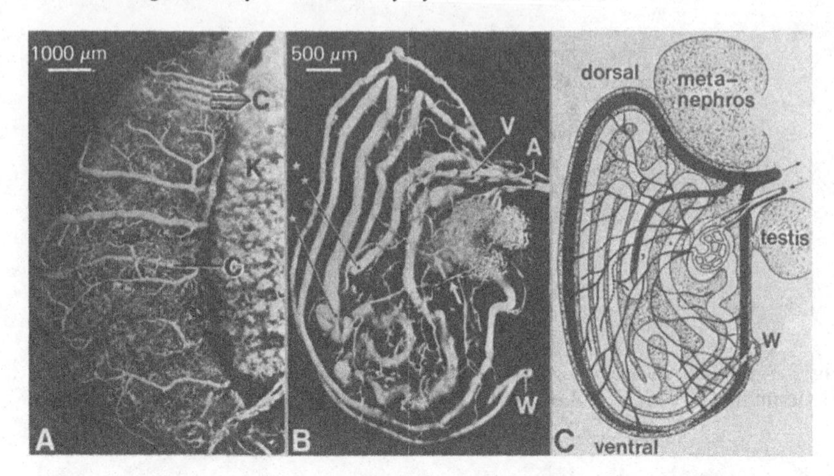

Fig. 1.8. Illustration of the vascularization of the mesonephric nephron (Tiedemann & Egerer, 1984; courtesy of Dr K. Tiedemann). A. Micrograph of a cast of microfil-injected mesonephric nephron (filled with microfil by rupture of glomerular capillaries, C). B and C. A glomerulus and its vascularization corresponding to A. The veins (V) and the arteries (A) can be followed from the hilus. The collecting tubule is drained by superficial veins while inner veins predominate in the dorsal part of the nephron. The peritubular capillaries originate from the efferent vessels of the glomeruli and supply wedge-shaped regions of the nephron. W, Wolffian duct; K, metanephros.

Croisille, 1976). These short branches of the Wolffian duct have apparently been invoked by an inductive stimulus from the developing nephrons (Croisille, 1976; Martin, 1976).

A detailed study of the vascularization of the mesonephric nephron has recently been performed by Tiedemann & Egerer (1984), who used vascular injections with microfil, followed by corrosion, and examination of the casts by scanning electron microscopy. In 5- to 8-cm pig embryos, the mesonephros is at the height of its development and becomes vascularized by 11 pairs of arteries from the abdominal aorta, each supporting up to 15 glomeruli. The detailed vascular pattern of a mature nephron may be seen in Fig. 1.8, which represents the first complete survey of the vascularization of a mammalian mesonephros.

The early condensation and S-shaped development and further maturation of the mesonephric nephron closely resemble that of the permanent kidney, the metanephros, but some definite, mainly ultrastructural differences between the two have been observed in a mature nephron. The juxtaglomerular apparatus is missing from the mesonephric tissue, and the mesangial component is conspicuous and rich in extracellular matrix. The fenestration of the capillaries is poor in the mesonephros, but compared to the metanephros there is a compensating abundance of capillaries. The glomerular basement membrane in

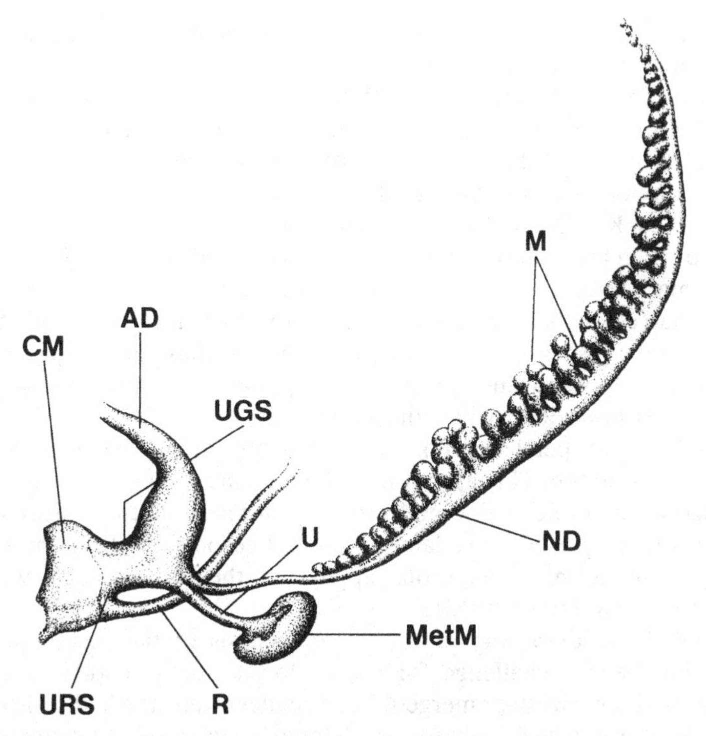

Fig. 1.9. Scheme of the urinary tract of an 8-mm human embryo (ovulation age 38 days) (after Winick & McCrory, 1968). M, metanephros; ND, nephric duct; MetM, metanephric mesenchyme; U, ureter; R, rectum; URS, urorectal septum; UGS, urogenital sinus; CM, cloacal membrane; AD, allantoic duct.

the mesonephros is frequently split into an epithelial and an endothelial component and is considered to be immature, resembling the glomerular basement membrane of an embryonic metanephric glomerulus (for reviews, see Berton, 1965; De Martino & Zamboni, 1966; Tiedemann & Egerer, 1984).

The metanephros

Introduction

The permanent kidney of the amniotes, the metanephros, is created by the two tissue components described above, namely the mesenchymal nephric cord (blastema) and the epithelial, originally mesoderm-derived Wolffian duct. From the most caudal portion of the Wolffian duct, now opening into the cloaca, a bud emerges and grows dorsally towards the caudal portion of the nephric cord (Fig. 1.9). The latter is seen as a dense condensate of mesenchymal cells into which the epithelial bud eventually grows. According to the classic view, the branches of the invading ureter should give rise to the entire epithelial component of the kidney, including the secretory nephron (Remak, 1855; Golgi, 1889), which would be analogous to the development of many glandular organs. It soon became evident, however, that both the epithelial and the mesenchymal components contribute to the formation of the nephron, as first suggested by Kupffer in 1865. Ultimately a third cell lineage becomes involved when the endothelial vasculature is created (Chapter 5).

In human embryos, development starts in the fifth week, when the ureteric bud emerges. In mouse this stage is reached on day 10. While the ureteric bud invades the mesenchyme and branches there, epithelial renal vesicles appear around its tips (the ampullae). Consequently, when new vesicles are added while the older ones continue to develop, a progressive, centripetal series of maturing nephrons may be detected in histological sections. This appositional development continues over an extended intrauterine period and proceeds in many species during the early postnatal period (Jokelainen, 1963; Leeson, 1971). In human embryos, the terminal nephrons appear in the thirty-second week (Osathanondh & Potter, 1963*c*).

The obvious complexity of the organogenesis of the metanephric kidney has been a challenge for embryologists for a century, and a voluminous literature has emerged. For a reader interested in the history of the field and in the background of today's views and controversial opinions, the reviews by Jokelainen (1963) and Kazimierczak (1971) are recommended.

While reviewing here the present concepts of kidney development, emphasis is laid on the organogenetic aspects of the process, and fewer details are included of the histogenesis and cytodifferentiation of the metanephric tissue components. I will focus on the development and ramification of the collecting system and on the early development of the secretory nephron. Little will be said of the guiding principles of these events, because the discussion of developmental mechanisms will be dealt with in Chapter 4.

The collecting system

Having invaded the mesenchymal blastema, the ureteric bud starts to divide dichotomously. The first branching pattern has been described in reconstruction models made from serial sections (Fig. 1.10). For an analysis of further ramification, other methods must be employed. On the basis of earlier microdissection efforts by Peter (1927), Oliver (1939) and by Darmady & Stranack (1957), an improved technique was developed by Osathanondh & Potter (1963*a*). The technique consists of acid maceration of whole embryonic kidneys or sections from later stages,

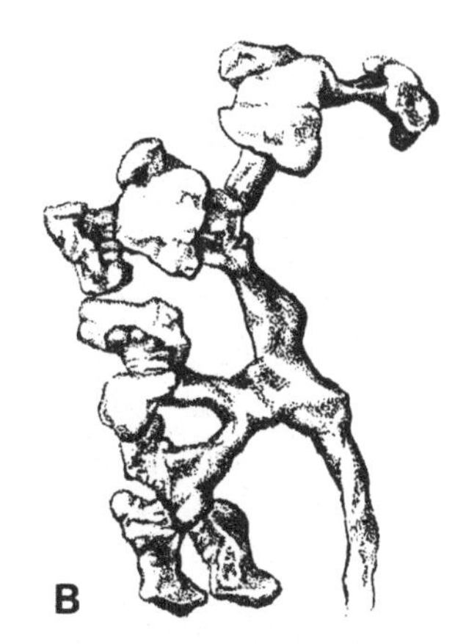

Fig. 1.10. Reconstruction models from serial sections of human embryonic kidney after the first branchings of the ureteric bud. A. Human embryo (16 mm). Wax reconstruction (after Huber, 1905). B. Human embryo (15 mm). Drawing from a photograph of a plastic reconstruction of Ludwig (1962).

followed by fractionated microdissection of individual nephrons to be examined microscopically (Fig. 1.11). In the following description of the development of the ureteric bud derivatives, the presentation is largely based on the classic reports by Osathanondh & Potter (1963*b*, *c*), who used this technique to examine human embryonic kidneys.

As shown by Osathanondh & Potter, both proliferation and branching of the ureteric bud are affected by local environmental factors, and the entire developmental chain is far more complicated than was previously thought. The dichotomous branching can be either symmetric or asymmetric. The former mode of ramification is the formation of two branches with similar sub-branching tendencies, whereas asymmetric dichotomy is the formation of two branches differing in their subsequent development: one might preferentially grow and elongate until a delayed branching occurs, while the other may enter the branching process immediately. Variations in the balance between elongation and ramification will consequently determine the ultimate architecture of the collecting system. This complicated series of events is the basis for the entire 'building plan' of the metanephric kidney, leading to the synchronous development of the three cell lineages – the epithelium itself, the mesoderm-derived nephrons, and the kidney vasculature (Chapters 3 and 5).

In the human kidney the first branching of the ureter in the mesenchymal blastema is dichotomous and symmetric and will create two branches that grow in opposite directions at an angle of 180°. The second branching is asymmetric and leads to the formation of one long and one short arm (Fig. 1.10). The development proceeds by further dichotomous

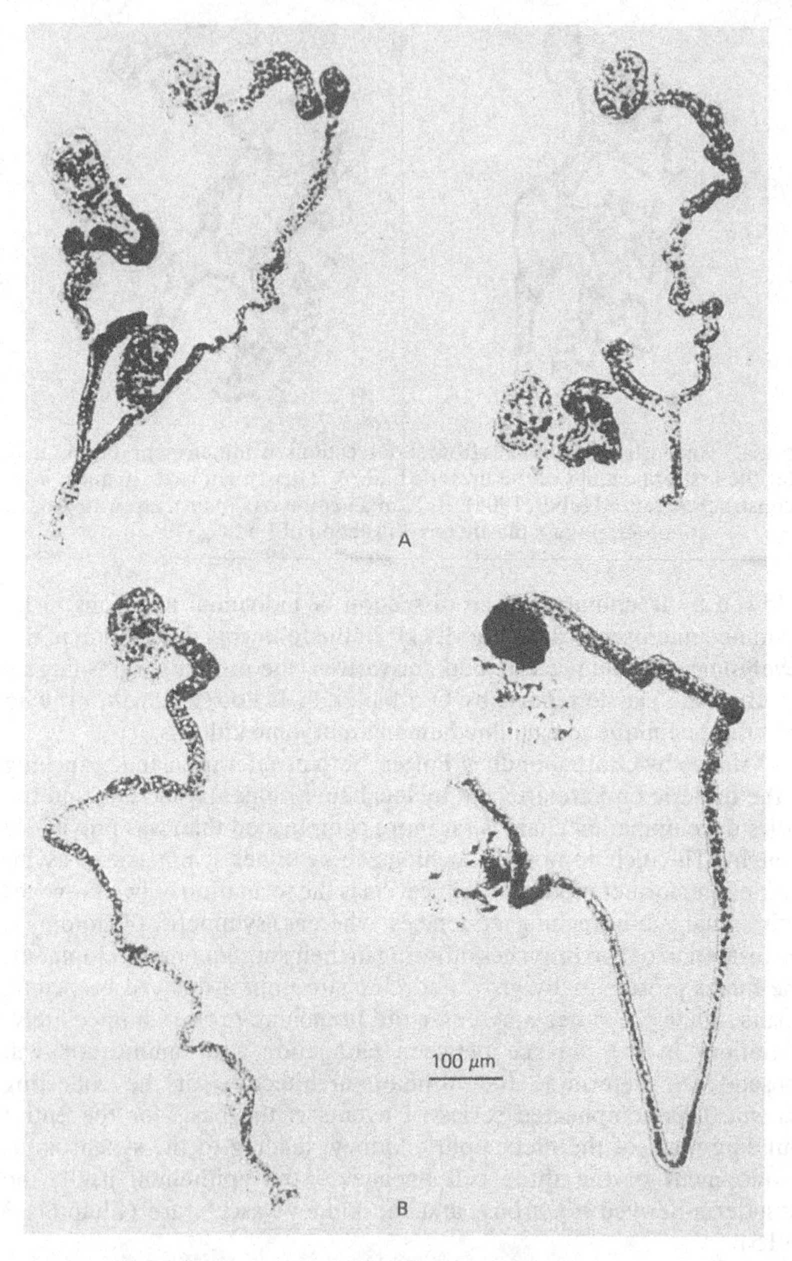

Fig. 1.11. Micrographs of human embryonic nephrons prepared by maceration and microdissection (Paatela, 1963; courtesy of Dr M. Viitanen). A. Premature child weighing 800 g. B. Premature child weighing 1200 g.

Fig. 1.12. Semi-schematic drawings of the early development of the human renal pelvis (after Osathanondh & Potter, 1963*b*).

branching with gradually diminishing angles. The first branches are soon taken up by the widening renal pelvis. They also form the calyses and papillae. The number of early branches swallowed by these formations vary in different parts of the kidney (Fig. 1.12). Due to this process, the branches that form the first-generation collecting ducts are 6 to 11 generations removed from the primary branches of the ureter.

To describe the ramification of the collecting system and its spatial development, Osathanondh & Potter (1963*c*) divided the process into four periods. The first is characterized by successive dichotomous branchings of the collecting system taking place at the widened tips, the ampullae (Fig. 1.13). Next to the ampullae, the first nephric vesicles, the prospective secretory nephrons are formed and soon join the collecting system. The early secretory nephron anlagen, attached to the collecting epithelium at the point in its zone of active growth, are carried deeper into the nephric mesenchyme by the growing collecting system.

The second period in the development of the collecting system is characterized by the formation of *arcades* (Fig. 1.13). Apparently the terminal branches of the collecting system (their ampullae) which have already completed their branching program retain their capacity to induce the formation of nephric vesicles. Thus, two nephrons may join the same ampulla, after which the older one shifts its point of attachment to the connection tubule of the younger. As the process goes on, new nephrons are formed and similarly joined to the same collecting duct until four to seven nephric anlagen contribute to one arcade (Fig. 1.14). A similar arcade formation has more recently been described in the rat, where the arcades consist of the three or four oldest juxtamedullary and midcortical nephrons (Neiss, 1982).

According to Osathanondh & Potter (1963*c*), the third period starts when the collecting duct grows past the attachment point of the arcade. The ampulla has ceased to branch, but, while growing, will induce a set of terminal nephrons, five to seven in the human kidney (Fig. 1.14) and two in the rat (Neiss, 1982). As a consequence of the above development, these subcapsular nephrons will join the collecting duct directly.

During the last period of development, the terminal ampullae

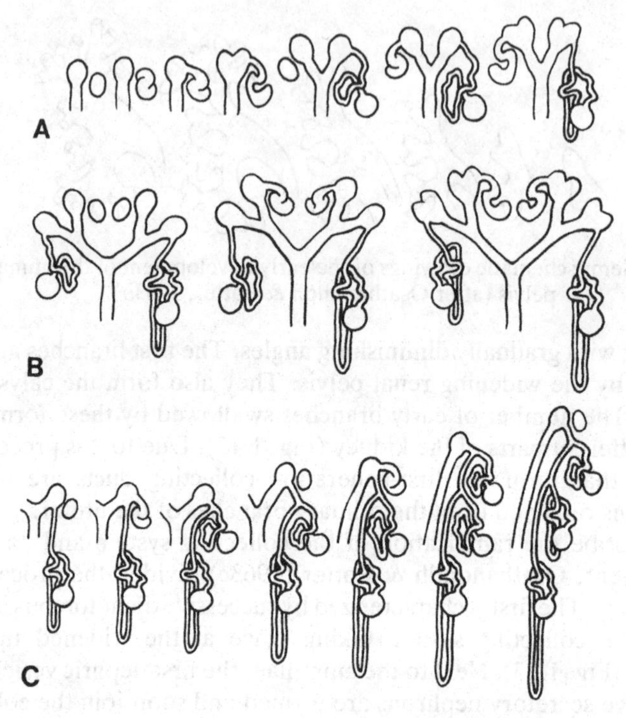

Fig. 1.13. Schematic presentation of the early development of the human nephron based on observations on nephrons isolated by microdissection (after Osathanondh & Potter, 1963*c*). A, B. Formation of the nephron anlage and the immature nephron. C. Formation of nephron arcades.

disappear and new nephric vesicles cease to form. This occurs at different stages in different species and may be delayed until the neonatal period.

The secretory nephron

The formation of the secretory component of the nephron, i.e. the epithelial part of the renal corpuscle, the proximal and the distal tubules and their derivatives, involves a unique conversion of loose mesenchyme into epithelium, which subsequently differentiates into several phenotypes. The early steps of this development were described in great detail at the beginning of this century by Herring (1900) and by Huber (1905). Their descriptions, based on light microscopy, and reconstruction models became the foundation of all subsequent investigations into early tubulogenesis. It soon became evident that the findings and views presented in these two classic papers were adequate in their essential aspects and that they were applicable to the developing metanephric kidney of all amniote species. The abundant literature that ensued from

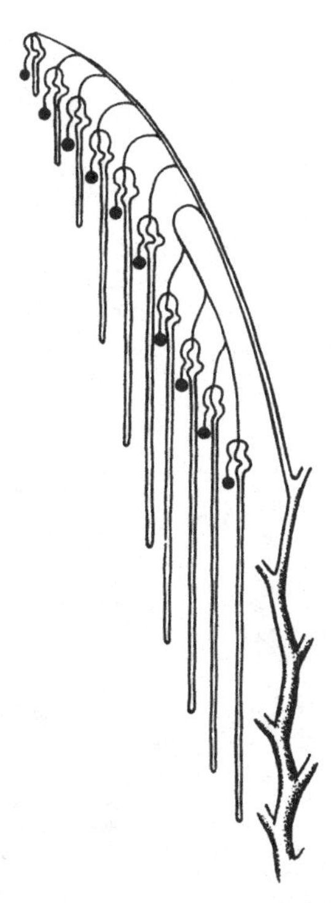

Fig. 1.14. Scheme of the arrangement of human nephrons at the time of birth. Four deeper constitute one arcade, while the four cortical nephrons are directly connected to the collecting duct (after Osathanondh & Potter, 1963*c*).

these papers has been reviewed by Jokelainen (1963), Potter (1965), Kazimierczak (1971), and Evan *et al.* (1984).

The first developmental changes in the metanephrogenic mesenchyme, following the invasion of the ureteric bud, have received little attention; most descriptions of tubulogenesis start from the formation of distinct, paired nephric 'vesicles' or pretubular aggregates. However, some definite changes precede the formation of these bodies, though they are difficult if not impossible to detect in fixed samples and tissue sections. The description below is therefore based almost exclusively on observations from time-lapse motion pictures obtained from mouse metanephric kidney anlagen cultured *in vitro* as three-dimensional whole-organ explants (Saxén *et al.*, 1965*a*).

Figure 1.15 presents prints of one such film and a corresponding camera lucida analysis of the early condensations and their relationships to the branches of the ureter described above. By the stage of the first, dichotomous and symmetric, branching of the ureteric bud, the mesenchyme has created a longitudinal condensate with somewhat indistinct borders. The primary, solitary condensate surrounds the entire ureteric

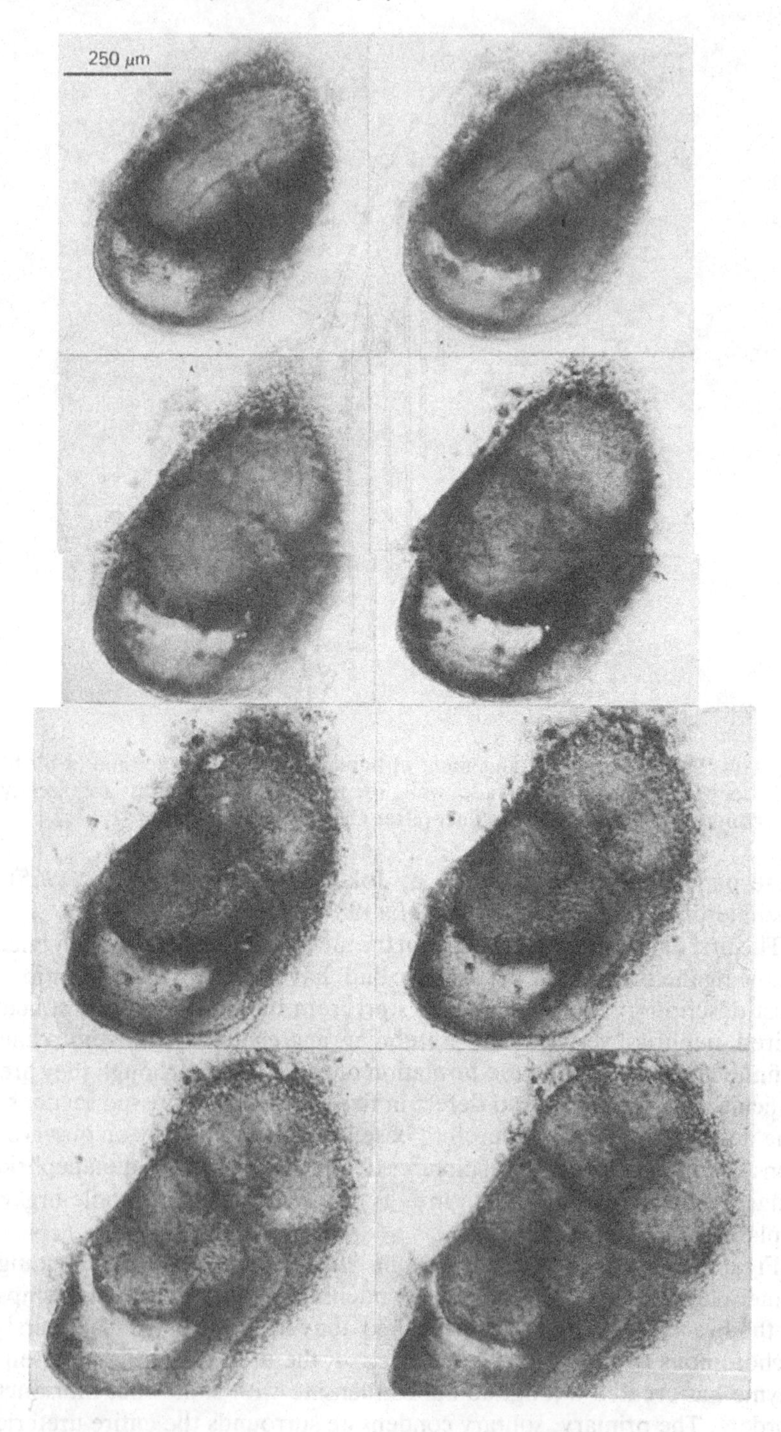
250 μm

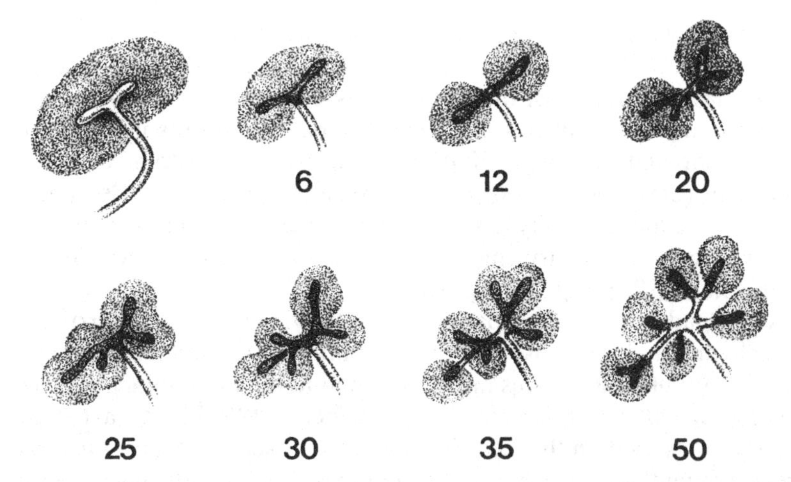

Fig. 1.15. Prints of a time-lapse motion picture on the early development of the mouse metanephric rudiment (opposite). Mesenchymal condensation and its subsequent splitting into secondary and tertiary condensates are seen in the prints and their relations to the ureteric bud and its branches in the camera-lucida drawings (above). (Saxén *et al.*, 1965*a*; Saxén & Wartiovaara, 1966). Numbers indicate time of cultivation in hours.

epithelium, as a 100- to 200-μm thick mantle growing both by cell multiplication and appositionally as new cells become trapped in its periphery. Interestingly, this maximal thickness, originally estimated from living kidney rudiments, corresponds well to the experimentally measured transmission distance of the inductive stimulus (Fig. 3.20, p. 83).

When the tips of the primary branches of the ureter grow apart at an angle of 180°, the solid condensate splits into two identical parts surrounding the growing tips. The second dichotomous ramification of the bud is asymmetric (like that described for human nephric anlagen in Fig. 1.10). It leads to the formation of secondary branches with a different mode of development and to the creation of an asymmetric configuration of the collecting duct system. This is followed by further fractionation of the mesenchymal condensate, as the splitting regularly occurs along the midline of two ureteric branches. With this continuous splitting and apparent 'packing' of cells within a condensate, the process ultimately leads to the formation of the distinct renal 'vesicles' which now become detectable in histological sections.

Some further observations made by the time-lapse cinematography and by immunohistology may provide the basis for speculative views on the mechanisms of this primary condensation process. The time-lapse motion pictures made on thick, three-dimensional metanephric anlagen do not allow precise analysis of the migration of single cells and their ultimate fate within the tissue, but show that cells trapped in a condensate move more slowly than those in the uncondensed part of the anlage. Yet, the cells have not fully lost their motility as a seemingly random movement still takes place over several cell layers. This results in constant displacement and exchange of position of the cells within an early condensate (Saxén *et al.*, 1965*a*). The exchange process has also been documented experimentally in various types of chick/quail recombinants by Armstrong & Armstrong (1973) and by Saxén & Karkinen-Jääskeläinen (1975) (Fig. 3.25, p. 86).

Factors leading to the cessation of cell motility, the condensation of the mesenchyme, and the subsequent fractionation of the primary condensate to pretubular aggregates might be found in the extracellular matrix directing cell behaviour in many tissues. In fact, in 1969 Linder made the interesting observation that certain connective tissue and foetal antigens were lost from the mesenchyme around the tips of the inducing ureteric bud during the early condensation phase. He suggested that this loss of the presumably extracellular material might be connected with the gradual disappearance of the intercellular spaces within the condensate. Later, Wartiovaara *et al.* (1976) showed that one such extracellular matrix component showing the pattern described by Linder was fibronectin, a glycoprotein associated with basement membranes in mature tissues (Vaheri & Mosher, 1978; Ruoslahti *et al.*, 1981). Another family of interstitial proteins frequently associated with morphogenetic events comprises certain types of collagens (Linsenmayer *et al.*, 1973; von der Mark & von der Mark, 1977; Thesleff *et al.*, 1979; Leivo *et al.*, 1980).

Changes in the distribution and expression of these proteins in the metanephric mesenchyme have been examined in more detail by Ekblom *et al.* (1981*b*) (Fig. 1.16). The uninduced metanephric mesenchyme uniformly expresses certain interstitial proteins such as collagen type I, collagen type III, and fibronectin. When, following an induction by the ureter, the mesenchyme forms the first condensates, these proteins seem to disappear, but are still found outside the condensed area. This loss (or reduction) of the extracellular material can well be thought to bring cells closer to each other, thus creating aggregates with restricted cell motility. The factors leading to the actual aggregation with cells attached to each other are not known, and several compounds have been suggested and shown to be present in the aggregates. One of these is laminin, a glycoprotein constituting a major component of the epithelial basement membranes, including those in the embryonic kidney (for a review, see Timpl & Martin, 1982). In an uninduced kidney mesenchyme, prior to any morphogenesis, immunohistology visualizes laminin as a diffuse, weak fluorescence throughout the blastema (Lehtonen *et al.*, unpublished results). Laminin fluorescence is intensified soon after induction, during early condensation of the mesenchyme around the tips

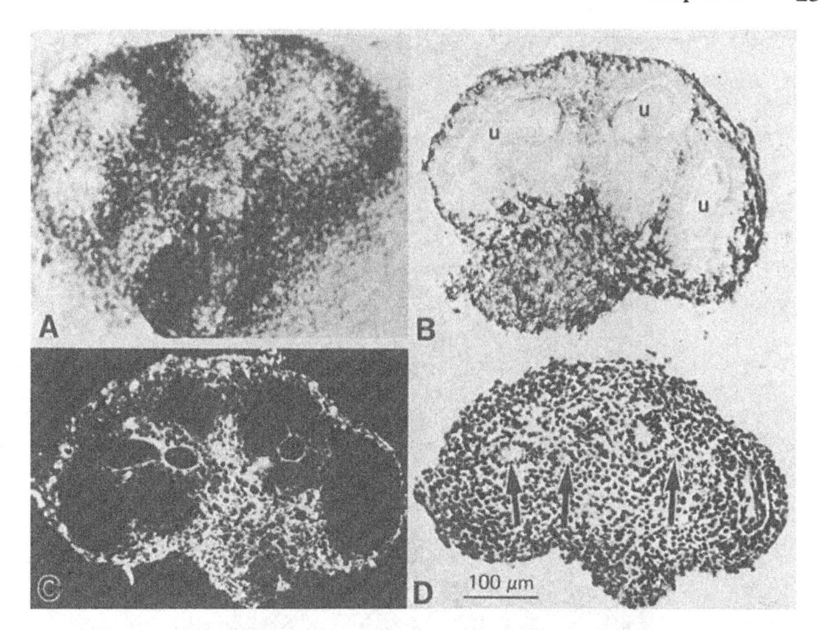

Fig. 1.16. Demonstration of the change in composition of the kidney extracellular matrix following induction and early mesenchymal condensation (Ekblom *et al.*, 1981*b*). A. Light micrograph of a living kidney from a 13-day mouse embryo. Note the mesenchymal condensates around the second-generation branches of the ureter. B. Micrograph of a section from a kidney corresponding to that in A. Immunoperoxidase staining with an antiserum against collagen type III shows the disappearance of the protein from the condensates surrounding the ureter (u). C. Immunofluorescence micrograph of a 13-day kidney demonstrating the distribution of collagen type III and its absence from the condensed areas. D. Section adjacent to C stained with haematoxylin and eosin. In routine histology, no condensation can yet be detected around the ureter (arrows).

of the ureter and may be now seen as distinct, large 'droplets', apparently both intracellularly and extracellularly (Fig. 1.17).

Later, following early renal vesicle formation and epithelialization of the mesenchymal cells, laminin becomes confined to the basement membrane. In the transfilter model-system, we could also show a temporal correlation between induction and enhanced laminin synthesis. These spatial and temporal correlations led us to suggest that laminin might contribute to the aggregation of the induced mesenchymal cells (Ekblom *et al.*, 1980*a*).

Other candidates might be found in the 'cell adhesion molecules' (CAMs) intensely studied in many tissues (for reviews, see Edelman, 1983, 1985; Thiery, 1984). One of these compounds, the N-CAM, has been shown to be expressed by newly aggregated mesenchymal cells of the mesonephric blastema (Fig. 1.18) and should be considered as possibly acting as an adhesive molecule in this event (Thiery *et al.*, 1982*b*,

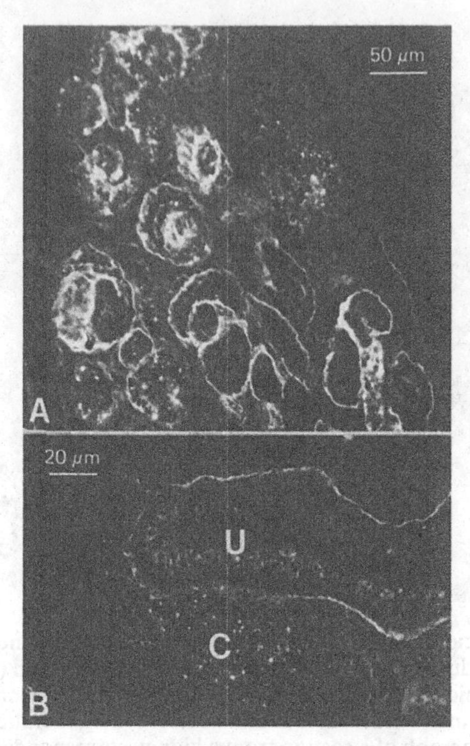

Fig. 1.17. Fluorescence micrographs of a developing mouse kidney treated with antiserum against laminin. A. A low-power view demonstrating the dotted appearance of laminin in the early condensations and its subsequent linear expression by the epithelial basement membrane (courtesy of Dr H. Sariola). B. A high-power micrograph illustrating the expression of laminin by the mesenchymal cells in the condensed area (C) and by the basement membrane of the inducing ureter (U) (Ekblom *et al.*, 1980*a*).

1984). An apparently closely related compound, the 120,000 M_r glycoprotein uvomorulin, has recently been detected in the metanephric kidney rudiment of the mouse (Vestweber *et al.*, 1985). It appears in the epithelializing cells at an early stage of condensation and might contribute to the aggregation of the mesenchymal cells. However, histogenesis could not be disrupted by the application of antibodies against uvomorulin in the model-system *in vitro*.

The gradual splitting of the primary mesenchymal condensate and aggregation of its cells ultimately result in the formation of the pretubular, paired anlagen in close vicinity of the ampullae of the ureter. Description of the further morphogenesis and histogenesis of the anlagen may be found in all embryology textbooks, though frequently somewhat inadequately visualized. They have been examined in great detail by light and electron microscopy, and for details the reader is referred to reviews

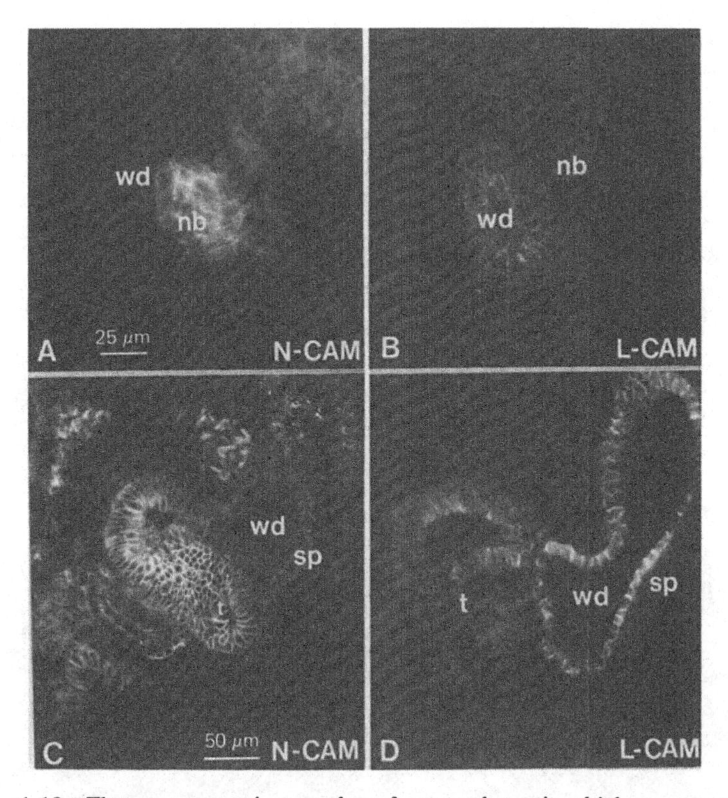

Fig. 1.18. Fluorescence micrographs of an embryonic chick mesonephros demonstrating the distribution of cell adhesion molecules (CAMs) in the aggregates and tubules (courtesy of Dr J.-P. Thiery). A, B. Sections through a mesonephros of a 25-somite embryo stained for N-CAM and L-CAM, respectively. The Wolffian duct-induced condensation (nb) expresses N-CAM but not L-CAM. Conversely, the Wolffian duct (wd) epithelium stains for L-CAM but not for N-CAM. C, D. Transverse sections through the mesonephros of a 35-somite embryo stained as in A and B. At this stage the mesonephric tubules (t) show a typical S-shape. While N-CAM is still expressed by the tubule cells, L-CAM appears mostly in the cells in close proximity to the Wolffian duct. sp, splachnic mesoderm.

by Jokelainen (1963), Potter (1965), Kazimierczak (1971), and Dørup & Maunsbach (1982).

Figure 1.19 illustrates the various steps of early tubulogenesis as seen in semi-thin sections by light microscopy. The sections were prepared from kidneys of 11- to 14-day mouse embryos. Figure 1.20 summarizes, in a slightly schematized way, our present concepts of the formation and shaping of the early secretory nephron. At the early 'vesicle' stage the polarized, somewhat elongated cells show a uniform ultrastructure and are vigorously proliferating, and the plane of division is radial to the cavity (Jokelainen, 1963). According to most descriptions, a central

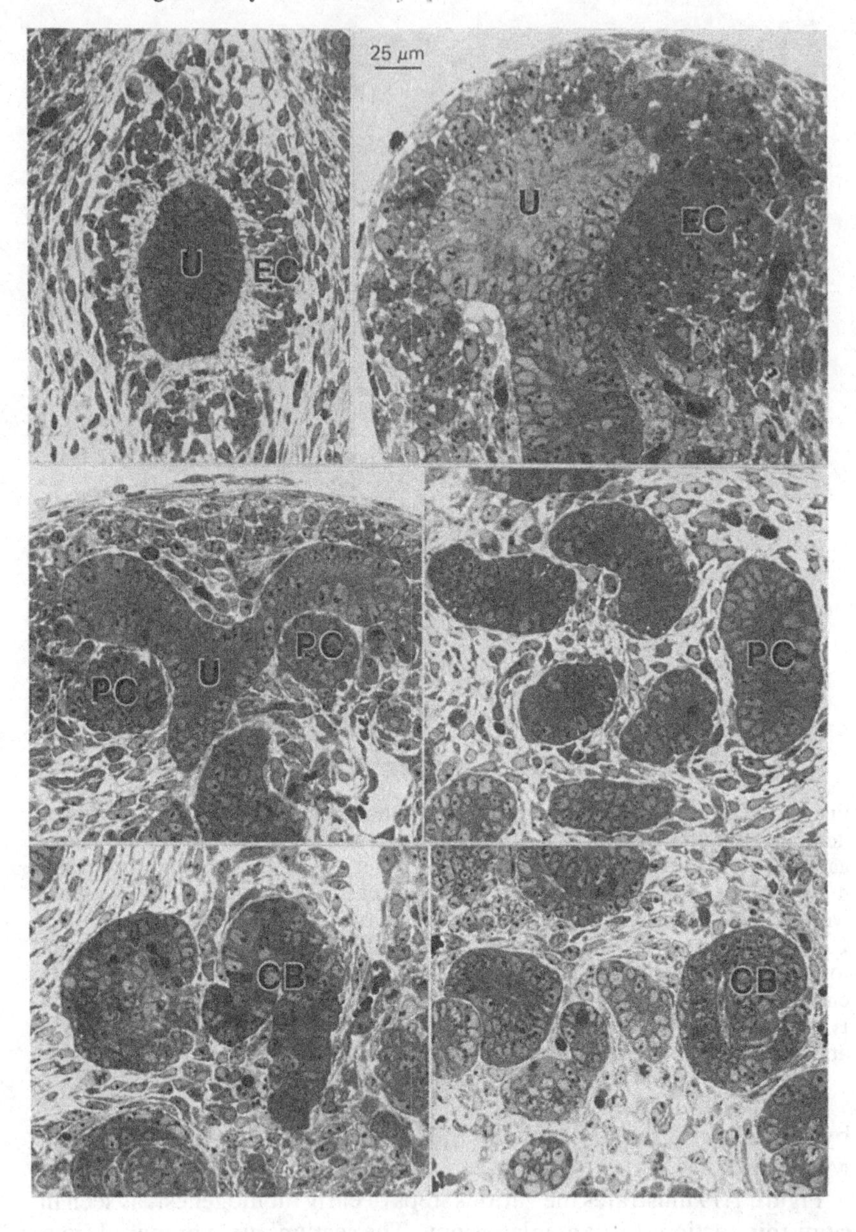
25 μm
U
EC
U
EC
PC
U
PC
PC
CB
CB

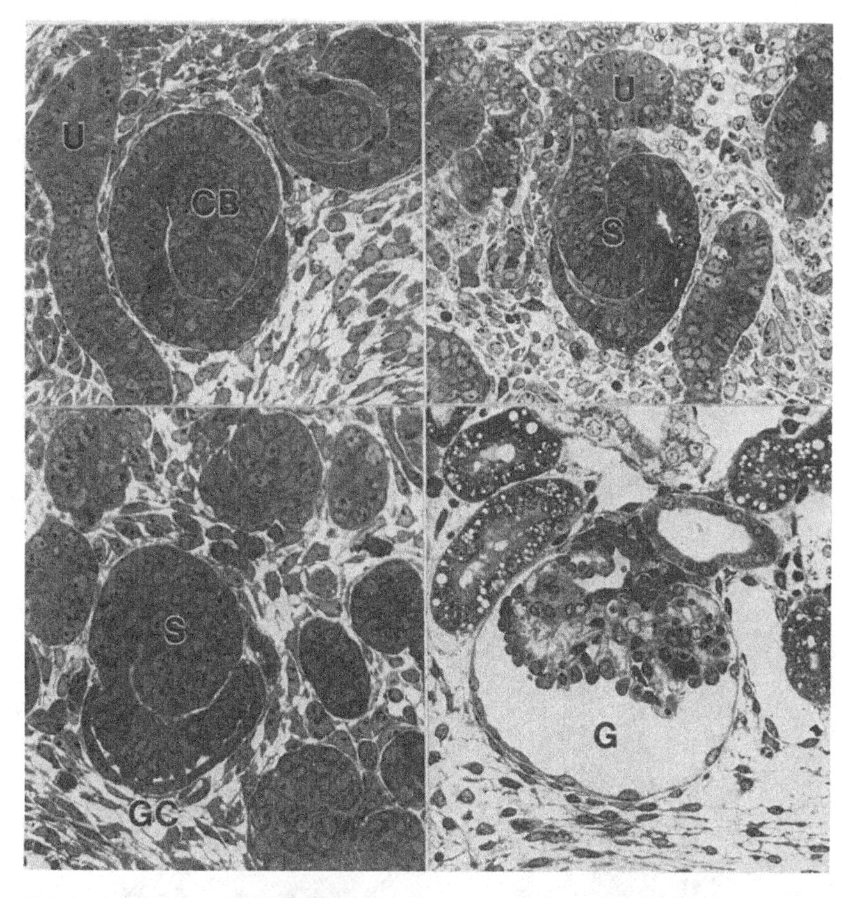

Fig. 1.19. A series of micrographs of semi-thin sections made from 11- to 14-day mouse kidneys. U, ureter; EC, early condensation; PC, pretubular condensation; CB, comma-shaped body; S, S-shaped body; GC, glomerular crevice with some endothelial cells; G, glomerulus.

lumen is found at this stage. But it was not found in our sections and it might be an artifact of fixation.

After this 'growth phase' the mitotic activity decreases, the anlage assumes a 'comma'-shape, and the first diversification of its cells is seen (Jokelainen, 1963; Dørup & Maunsbach, 1982; Fig. 1.21). Cells farthest from the collecting duct (at the proximal pole) become elongated, their nuclei change position, and some cells become funnel shaped. This is the site of subsequent slit formation, the first sign of development of the glomerular crevice. While this process is going on, a second slit is gradually formed at the opposite, distal pole, and the tubular anlage now assumes the well-known S-shape and subsequently joins the collecting system.

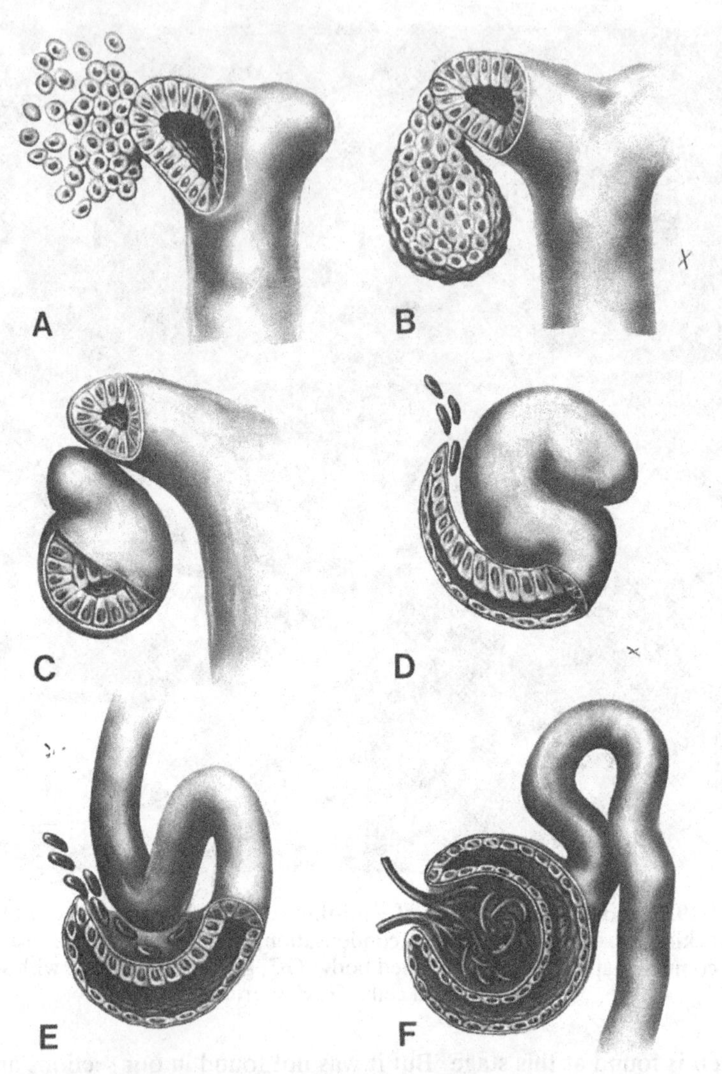

Fig. 1.20. A to F. Semi-schematic illustration of the early development of the nephron from condensation to the **S**-shaped body (after Saxén, 1984*a*).

At the **S**-shape stage, cell diversification becomes apparent. The cells of the future Bowman's capsule become flattened (Dørup & Maunsbach, 1982), and the presumptive podocytes differ from the rest of the cells by their lectin-binding capacity. As shown in human embryonic kidneys, these cells bind three lectins (*Maclura pomifera* agglutinin, wheat-germ agglutinin and peanut agglutinin) not bound to other cells within the **S**-shaped body (Holthöfer & Virtanen, 1986; Fig. 1.22). In the embryonic mouse kidney, some regional specificity of the glomerular portion of the

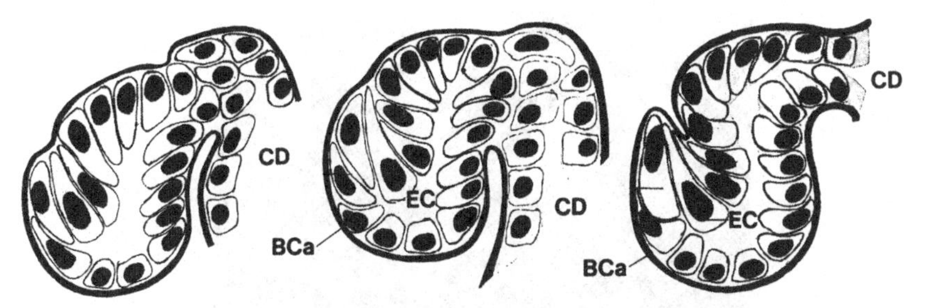

Fig. 1.21. Semi-schematic drawings from thin sections of rat kidney, illustrating the histogenesis of the early renal vesicle (Jokelainen, 1963). CD, collecting duct; BCa, Bowman's capsule; EC, epithelial cells with displaced nuclei.

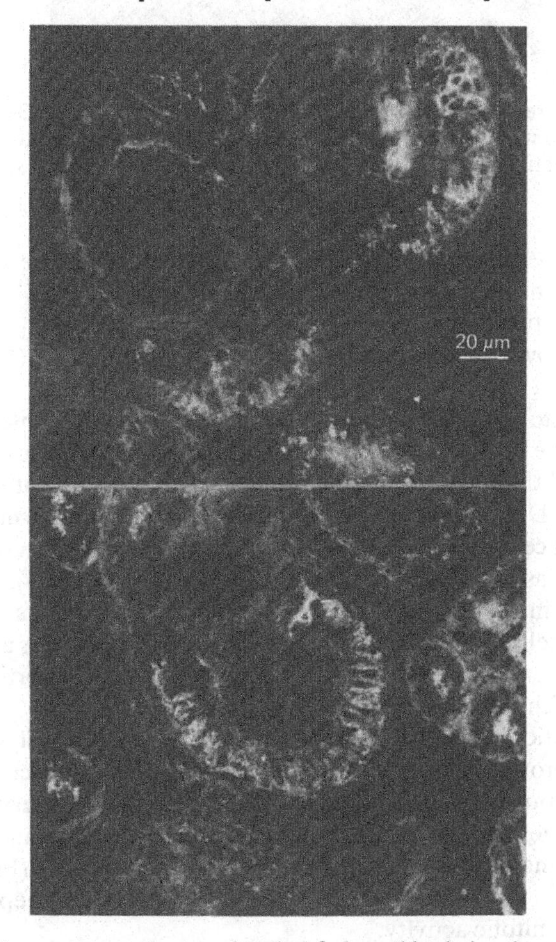

Fig. 1.22. Fluorescence micrographs of an S-shaped body in a human embryonic kidney. The sections have been treated with fluorescein-conjugated *Maclura pomifera* lectin, which selectively binds to the presumptive podocytes (Holthöfer & Virtanen, 1985; courtesy of Dr H. Holthöfer).

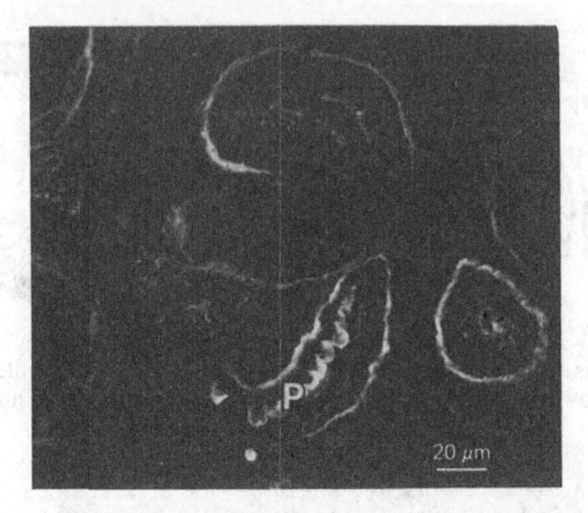

Fig. 1.23. Fluorescence micrograph of an S-shaped body in a 13-day mouse kidney. Treated with fluorochrome-conjugated lectin peanut agglutinin decorating the presumptive podocytes (P) and the basement membrane (Laitinen *et al.*, 1986; courtesy of Dr L. Laitinen).

nephron can be detected soon after the S-shaped state. The podocytes show a transient binding of peanut agglutinin and soya bean agglutinin, but during maturation these sites are apparently covered by sialic acid moieties as shown by their binding of wheat-germ agglutinin (Laitinen *et al.*, 1986; Fig. 1.23).

Immunohistological exploration of the expression of intermediate filaments of the cytoskeleton at the early stages of nephron development have revealed that there is already some cell heterogeneity at the late S-shaped stage. During the formation of the renal vesicle in the mouse, the mesenchymal cells lose their vimentin-type fibres and soon start expressing cytokeratins in a fibrillar manner (Lehtonen *et al.*, 1985). At the S-shaped stage in the human kidney, a third cell type appears. Both the presumptive cells of the Bowman's capsule and the podocytes are devoid of cytokeratin, which at this stage appears in the tubular portion of the nephron (Holthöfer *et al.*, 1984; Fig. 1.24).

The formation of the S-shaped body is generally attributed to a 'differential growth' of the renal vesicle, and this view is reflected in most of the illustrations in embryology textbooks. Our direct observations by time-lapse cinematography do not sustain this concept, but strongly suggest a formation of the slit by cell detachment *in situ* (Fig. 1.25). Subsequently, of course, the different segments of the nephron are elongated by mitotic activity.

Electron microscopy of the early renal vesicles both in whole kidneys and in experimentally induced mesenchymes has shown that a basement membrane accumulates around the aggregate soon after its formation

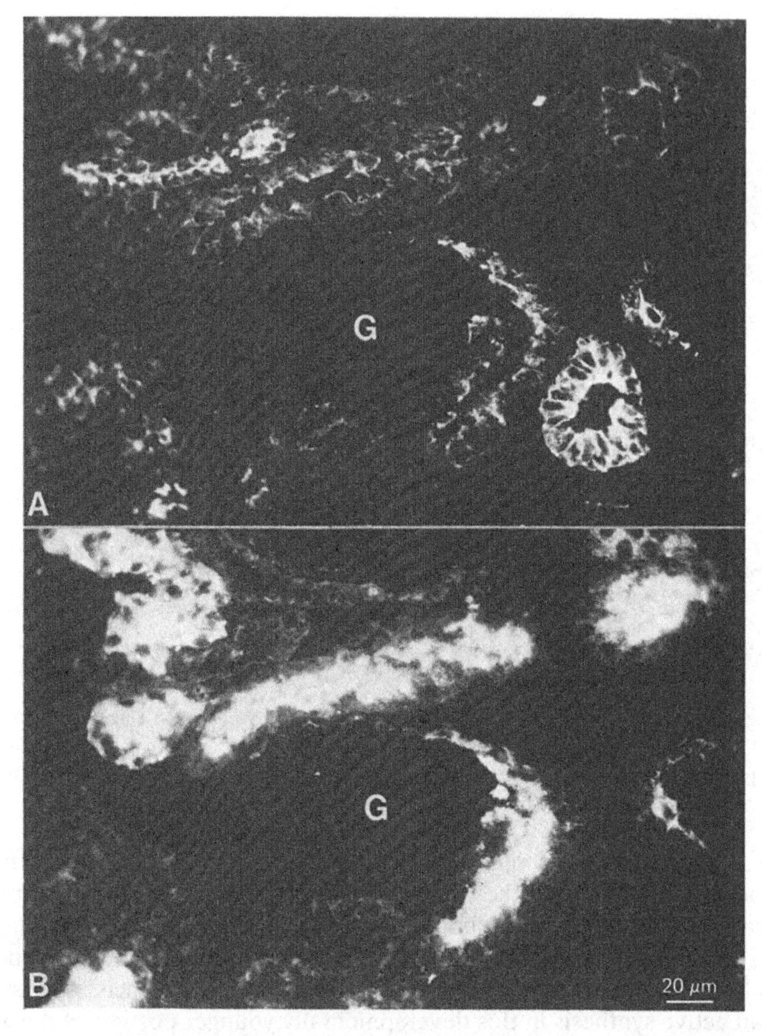

Fig. 1.24. Fluorescence micrographs showing the distribution of cytokeratin (A) and the brush-border (BB)-antigen (B) in human foetal nephrons at the S-shaped stage. Visualized by double staining in immunohistology (Holthöfer *et al.*, 1984; courtesy of Dr H. Holthöfer). The glomerulus (G) is devoid of cytokeratin. The proximal tubules express both cytokeratin and the BB-antigen.

and is completed at early S-shape phase, when it also extends into the slit formation (Jokelainen, 1963; Wartiovaara, 1966*b*; Abrahamson, 1985). This can be well illustrated by immunohistology with antibodies against the main constituents of the basement membrane, collagen type IV, laminin, and heparan sulphate proteoglycan. The co-distribution of these three is complete, and we may therefore follow just one, laminin. In the early condensates, laminin appears as randomly distributed granules

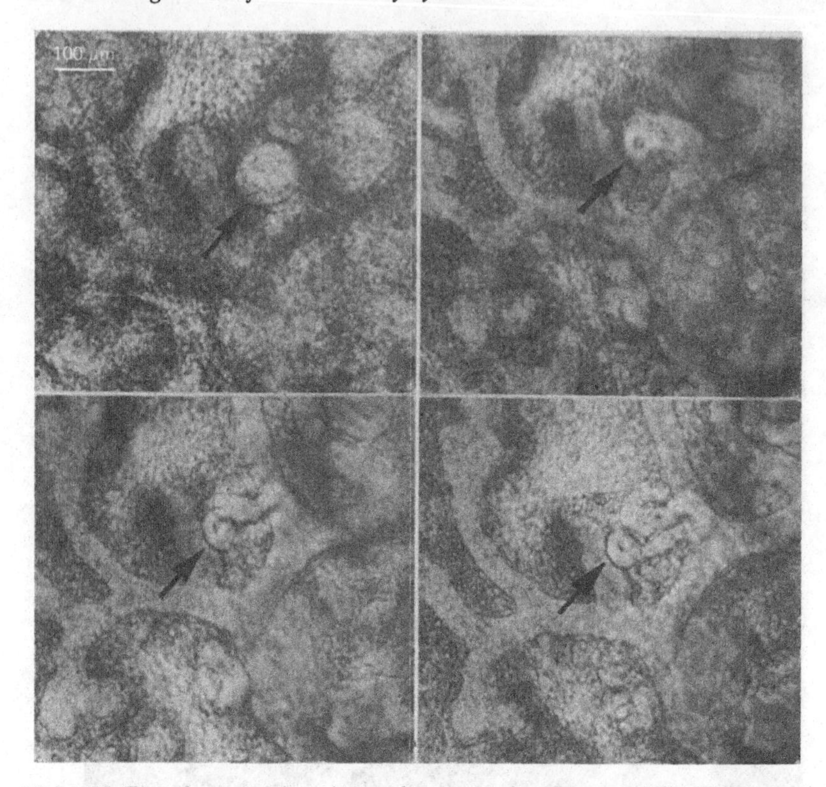

Fig. 1.25. Time-lapse motion picture demonstrating the remodelling *in situ* of the nephric vesicle into an **S**-shaped body (arrows) (Saxén *et al.*, 1965*a*).

(Fig. 1.17), but by the comma-shape stage it is confined to the basement membrane area around the anlage. New protein is laid down during growth and reshaping of the body, and during the **S**-phase a continuous layer of laminin is seen around the main portion of the anlage. At the distal pole, however, the laminin is still granular in appearance, suggesting an active synthesis in this developmentally younger portion of the **S**-shaped body (Ekblom, 1981*a*). As in many epithelial tissues, the close temporal and spatial correlation of appearance of the basement membrane and the epithelial shaping of cells suggests a causal relation. Consequently, it has been suggested repeatedly that the basement membrane may act as a scaffold for the epithelial cells, which become attached to the basement membrane and polarize (see, for example, Bernfield *et al.*, 1972, 1984*a*, *b*; Ekblom, 1984; Hay, 1984).

The mechanism of the remodelling of the **S**-shaped body that involves the formation of a double slit is an intriguing, unsolved problem. Such events have frequently been attributed to local changes in cell-to-cell adhesiveness or affinity (Holtfreter, 1939; Gustafsson & Wolpert, 1963; Kiremidjian & Kopac, 1972; Hilfer & Hilfer, 1983; Mittenthal & Mazo,

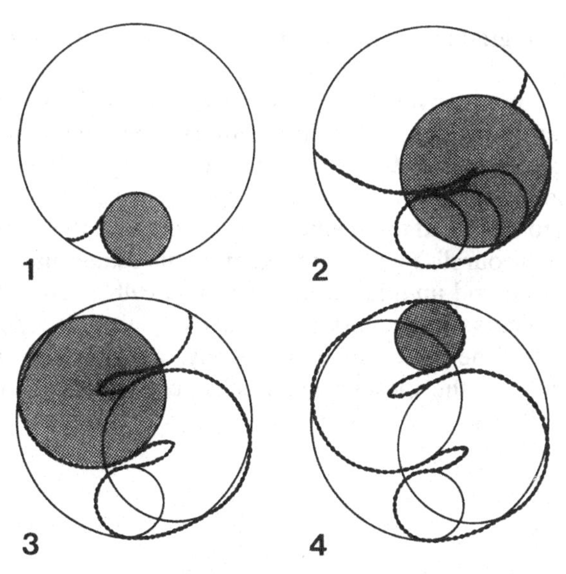

Fig. 1.26. Scheme of the formation of the S-shaped body according to a hypothesis of a postinductory, gradually increasing, intercellular adhesiveness and an adhesive gradient (after Saxén, 1970*b*).

1983). We have presented a hypothesis on the development of the S-shaped body, based on an assumption of a postinductory, gradually increasing intercellular affinity of the pretubular mesenchymal cells (Saxén & Wartiovaara, 1966; Saxén, 1970*b*): due to induction by the tip of the ureter growing into the mesenchyme and to the spatial organization of the two tissue components (Fig. 1.20), it can be assumed that the cells of the lower (proximal) pole have received the trigger earlier than those at the upper (distal) pole next to the collecting duct. If induction leads to a gradually increasing intercellular affinity (for which indirect evidence is presented below), the first induced cells at the proximal pole should be more adhesive than those at the last induced distal pole, and, in fact, an adhesive gradient along the proximo-distal axis might have developed. The following sequence of events may be predicted (Fig. 1.26): due to increased mutual adhesiveness, cells now attached to the basement membrane tend to maximize their mutual contact surface, which is seen as elongation of the cells. A second consequence would be that cell colonies would tend to round up (nephric vesicle) and become separated from less adhesive cells. As the most adhesive cells in the S-shaped body are those at the proximal pole, one might expect this process to be initiated there, and the histogenesis described is suggestive of such a phenomenon. Cells become detached from cells of less affinity at the midportion of the body, but remain in contact with their immediate neighbours within the gradient, as differences in affinity are minimal there. Hence, a gradual 'pinching off' of the S-shaped cell population occurs *in situ*.

Needless to say, the hypothesis is unproved and might well remain as such. However, in addition to the direct observations on the histogenesis and morphogenesis of the S-shaped body supporting such a view, some indirect evidence for increased adhesiveness is available. Induction, indeed, leads to increased intercellular adhesiveness of the mesenchymal cells, as shown by our disaggregation experiments comparing the dissociability of induced and uninduced mesenchymal cells (Saxén & Wartiovaara, 1966). The same increased adhesivity might be deduced from observations by time-lapse cinephotography, according to which the random motility of the mesenchymal cells ceases during aggregation (Saxén *et al.*, 1965*a*).

2

Experimental methods to study kidney development

Introduction

Most of our knowledge on the morphogenesis of the vertebrate kidney, reviewed in the previous chapter, has been gained from lengthy, descriptive studies on fixed material, with the exception of some experimental studies mentioned on p. 20. When classic, descriptive embryology shifted to developmental biology, focusing attention on the dynamics of development and on the control mechanisms of embryogenesis, an experimental approach became necessary. Analysis of development could now be implemented by experimental interference with various developmental events either *in vivo* or in various grafting systems. *In vivo*, the operations included destruction of specific tissues and organ anlagen or their transplantation into heterotopic sites isolated from neighbouring tissues. A complete separation from organismal influences could be achieved either by grafting the tissues into various hosts or by cultivating them *in vitro* under strictly controlled conditions. Since understanding of many present ideas of nephrogenesis – as well as appreciation of the lack of them – requires some knowledge of the basic techniques, they will be reviewed briefly in this chapter. The present grafting and culture techniques are emphasized, especially those devised primarily for studies of kidney development. More details of such experiments will be presented later, when their results and conclusions are discussed.

Early experiments *in vivo*

At the beginning of this century, embryologists who were exploring the embryogenesis of lower vertebrates developed many of the basic microsurgical techniques that are still in use. Experiments with amphibian embryos that were easily available and could be manipulated included transplantation of tissues, destruction of certain anlagen, and cell labelling. Through them we can understand the dynamics of early embryogenesis and the control mechanisms involved (for reviews, see Spemann, 1936; Saxén & Toivonen, 1962). These early experiments also included some investigations on early nephrogenesis (p. 3).

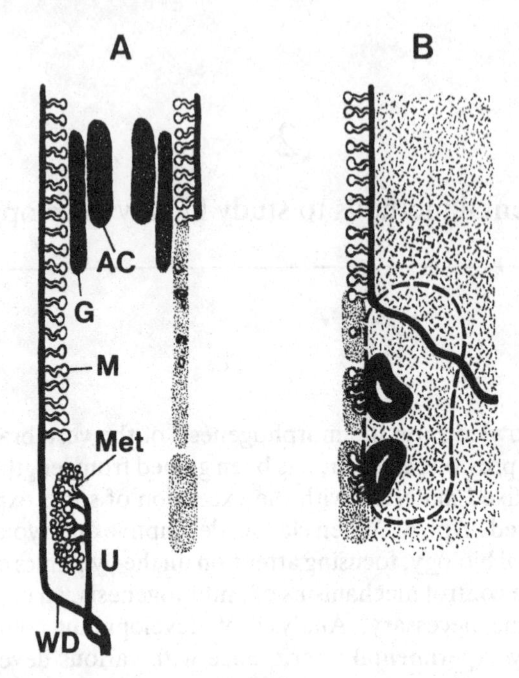

Fig. 2.1. Experimental scheme and results of microsurgery *in vivo* by Gruenwald (1942). AC, adrenal cortex; G, gonad; WD, Wolffian duct; M, mesonephros; Met, metanephros; U, ureter. A. Normal anatomy of the urogenital tract of a chick embryo (left), and the effect of destruction of the growing end of the Wolffian duct (right). B. Experimental introduction of non-nephrogenic tissue between the Wolffian duct and the nephric mesenchyme. The growth of the duct is distorted, and the metanephric tubules are formed only at sites where the transplanted neural tissue makes contact with the nephric mesenchyme.

The first indication of an inductive tissue interaction guiding the development of the nephron came from such experiments *in vivo*. In 1927, Boyden managed to produce, experimentally, an obstruction of the chick mesonephric (Wolffian) duct with the consequence that the nephric blastema did not differentiate (Table 3.1, p. 53). This approach was followed by three fine experiments by Gruenwald (1937, 1942, 1943). In the first, he confirmed Boyden's results and conclusions by destroying the growing end of the Wolffian duct and subsequently observing no differentiation of the nephric mesenchyme. In the second series of experiments, Gruenwald (1942) prevented contact between the Wolffian duct and the nephric mesenchyme by transplanting non-nephrogenic embryonic tissues between them. As a result, the Wolffian duct deviated from its normal route of growth and did not make contact with its mesenchymal counterpart (Fig. 2.1A). Consequently, no nephric tubules developed. Nevertheless, Gruenwald made the interesting observation that when the transplant carried neural tissues, tubules developed at the sites where the transplant made contact with the nephric mesenchyme (Fig. 2.1B). This finding led Gruenwald (1943) to his third experiment *in vivo*: embryonic nervous tissue was grafted so as to make contact with the mesonephric and the metanephric mesenchymes, with the expected result of kidney tubule formation at the contact sites. Ten years later these observations led to the classic experiments *in vitro* on kidney tubule induction (Chapter 3).

Grafting techniques

Chorioallantoic grafts

The chorioallantoic grafting technique was developed at the beginning of the century to observe the development of isolated embryonic tissues and organ rudiments. The first to use this technique for exploring nephrogenesis was Atterbury in 1923. She dissected metanephric anlagen from seven-day chick embryos and grafted the whole anlage or its fragments onto the chorioallantoic membrane (CAM) of embryos of the same age. Good survival and growth were obtained in these grafts during the seven-day grafting period. Differentiation was reported to be 'exactly similar' to that occurring within the embryo, and the differentiated structures included epithelial tubules and vascularized glomeruli. The author, quite properly, concluded that the vascular component of the glomerulus was derived from the allantoic capillaries.

To examine the developmental potentials of the various levels of early chick embryos, Seevers (1932) grafted these tissues onto CAM, applying a technique described by Willier (1924). Grafts from the caudal level of embryos at the thirty-first somite stage and onwards developed complete kidneys, whereas similar grafts from younger embryos never showed kidney structures after the 9- to 10-day grafting. As the above critical stage corresponds to the stage of nephrogenesis where the Wolffian duct is brought into contact with the nephric mesenchyme, Seevers concluded that 'the differentiation of the metanephric blastema is dependent on ureter formation'.

The most extensive series of grafting of chick kidneys onto CAM have been performed by the French school led by Etiènne Wolff (for a review, see Wolff *et al.*, 1969). The classic grafting technique was somewhat modified, with very good results (Calame, 1961): on day 3 of incubation, 1 to 2 ml of albumin is sucked from the small end of the egg, then a window is cut over the embryo that is now separated from the shell membranes. The window is sealed and re-opened after an additional 2- to 4-day incubation. The allantoic vesicle is easily accessible without damage to the embryo. The graft is introduced into a slit, cut at the site of an anastomosis of the CAM vessels and kept in place by the edges of the

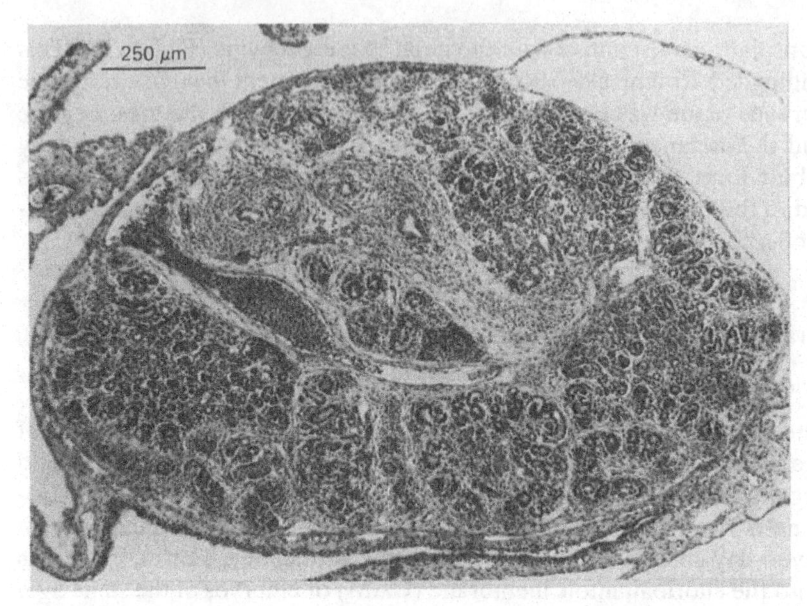

Fig. 2.2. Micrograph of an 11-day mouse metanephric rudiment grafted onto quail chorioallantoic membrane for six days. Abundant vasculature and advanced nephric structures may be seen (Ekblom *et al.*, 1982).

slit. This technique, allowing good growth and differentiation of the graft, was employed in an extensive experimental series where the epithelial and mesenchymal components of the kidney were grafted either separately or after combination to various heterotypic tissues. Such series could be done by combining the grafting technique to a short-term organotypic culture, i.e. the tissue combinants were first cultivated *in vitro* for a short period and after fusion grafted onto CAM as described above (Bishop-Calame, 1965*a*, *b*, 1966). Some of the results will be described in the next chapter.

Mouse kidney anlagen and their components were grafted in this way by Grobstein & Parker (1958), who used precultivation *in vitro* to produce large tissue masses of separated metanephric mesenchymes. Preminger *et al.* (1980, 1981) used 10- to 12-day mouse kidney anlagen and grafted them on 8- to 10-day chick CAM after gentle abrasion of the membranes between two branching arteries. To keep the graft in place, they covered it with a polyethylene film fixed with a silicon rubber ring. Good survival, angiogenesis and formation of rather advanced nephrons including glomeruli were obtained in these conditions after five to 10 days. This result is in accord with our experience from grafts of undifferentiated mouse kidney rudiments on quail CAM (Fig. 2.2). An advantage of this host is the quail 'nuclear marker' that allows the identification of quail cells within the mouse graft (Chapter 5; Sariola *et al.*, 1983; Saxén, 1984*b*).

Allogeneic grafts

Efforts have been made to graft embryonic and neonatal kidneys or their fragments into various sites of allogeneic hosts. Waterman (1940) grafted rabbit mesonephric tissue onto the omentum and reported that previously formed tubules survived well up to a total tissue age of 21 to 43 days. Rat mesonephric anlagen at different stages of development were grafted intraocularly by Runner (1946), who obtained tubule formation only occasionally, when the grafting was done prior to the initiation of morphogenesis *in vivo*. After formation of the tubules *in vivo*, the grafts continued to differentiate, and the tubules survived much longer than usual under these conditions.

Grobstein & Parker (1958) grafted mouse metanephric kidneys and their components onto such sites as the anterior chamber of the eye, the brain, and the subcutaneous tissue of adult mice. Whole kidney rudiments and isolated metanephric mesenchymes survived well in the eye and the brain but poorly in the subcutaneous sites. Interestingly, the isolated mesenchymes that failed to form epithelial tubules *in vitro* did so in the heterotopic sites *in vivo* – in both the eye and the brain. For chorioallantoic and intracoelomic grafts, two to eight mesenchymes were fused by a short precultivation *in vitro*. On CAM these grafts did not differentiate, whereas the intracoelomic explants showed abundant tubules. Whole-kidney rudiments grafted similarly in the chick coelom differentiated well and formed glomeruli.

Renal tissues from foetal and neonatal rats were similarly grafted onto the host's subcutaneous tissue by Barakat & Harrison (1971). The relatively large grafts usually showed a central, ischemic necrosis, but the peripheral parts became vascularized and survived, and the authors reported tubule formation in such grafts.

Important results have been obtained by isotopic and isochronic transplantation of fragments of quail neural tube into chick embryos. By this technique, Le Douarin & Teillet (1974) demonstrated colonization of early kidney blastemas by cells of neural crest origin.

Experiments *in vitro*

In addition to grafting techniques, various methods *in vitro* have proved useful for exploring single events of nephrogenesis at the molecular, cellular and supracellular levels. Cell and tissue culture methods combined with modern analytical tools allow precise analysis of various isolated events in development, but they are also likely to create artificial results that are frequently confused with true developmental processes. The limitations, pitfalls, and drawbacks of these techniques will not be discussed thoroughly here, as they are dealt with in handbooks and manuals on tissue culture.

Cell culture

Kidney cells can be studied *in vitro* either in continuous, established lines of normal and neoplastic cells or in primary cultures. Established cell lines provide good model-systems for studying problems in general cell and molecular biology. They may be less useful when events are explored that are specific to the kidney and its development. Then primary cultures of kidney epithelial cells from various parts of the nephron as well as mesenchymal and endothelial cells of the kidney have proved more rewarding.

Glomerular cells from human and animal kidneys have been widely studied (for reviews see Striker *et al.*, 1980; Kreisberg & Karnovsky, 1983; Nörgaard, 1983). Glomeruli are isolated from fragments of the kidney cortex by passing small tissue fragments through a stainless-steel screen or through a Teflon net with pore sizes from 120 to 250 μm, depending on the species. The glomeruli are purified by centrifugation and further filtration, followed by seeding onto culture dishes or dissociated to obtain single cell suspensions for ultimate plating. In either case, the resulting cell cultures consist of at least three different cell types. For further purification, many cloning or selection processes have been devised (Foidart *et al.*, 1979; Kreisberg & Karnovsky, 1983). Various pure cultures of glomerular epithelial cells, mesangial cells, and vascular endothelial cells have been obtained.

Kreisberg & Karnovsky (1983) have summarized information gained from studies on glomerular cell cultures. This includes data on synthetic and metabolic activities of the glomerular epithelial cells, such as synthesis of the molecules contributing to the glomerular basement membrane and synthesis of prostaglandins. Furthermore, at least two types of mesangial cells have been distinguished and characterized, and their antigenic properties mapped out. As for organogenesis, such studies and observations are of great value, but they should be complemented with results from somewhat more complicated, preferentially three-dimensional model-systems.

Kidney tubule cells have also been cultured, though their exact origin is still difficult to determine. By trypsin digestion and repeated sievings of kidney fragments, Cade-Treyer (1972) and Cade-Treyer & Tsuji (1975) obtained suspensions of glomeruli and of tubules (Fig. 2.3), from which they prepared primary cell cultures. The method of Oberley & Steinert

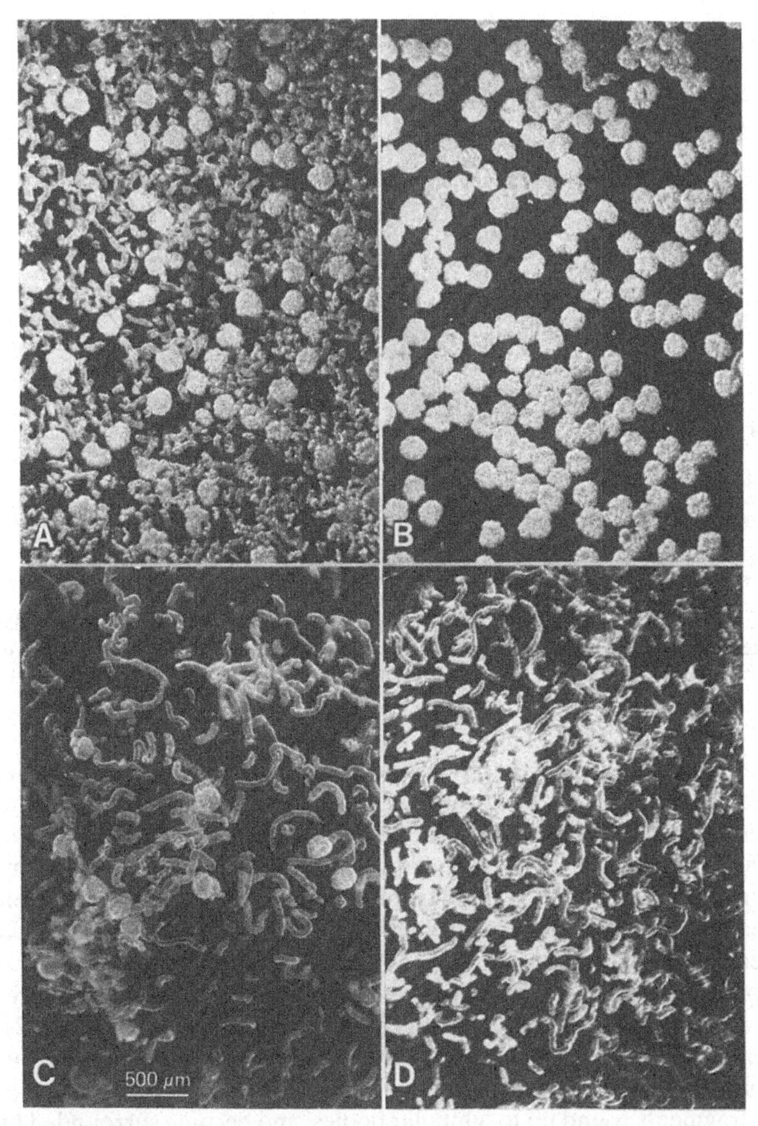

Fig. 2.3. Illustration of the separation and purification of glomeruli and of tubules from calf kidney (Cade-Treyer, 1972; courtesy of Dr D. Cade-Treyer). A. Trypsinized sample from an adult kidney, after the first sieving and removal of free cells, contains both glomeruli and tubules. B. Pure glomeruli after a second sieving. C. Trypsinized suspension of foetal calf kidney cortex. D. Preparation of pure tubules.

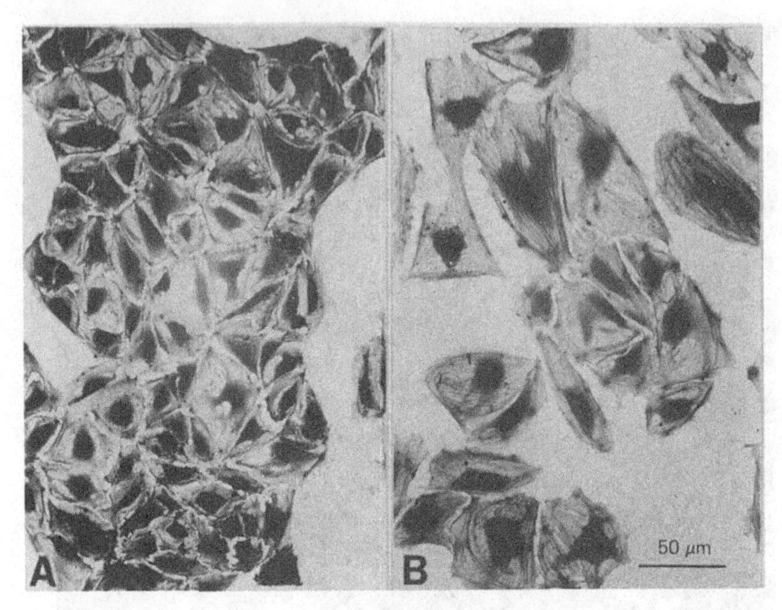

Fig. 2.4. Micrograph of renal cortical cells in culture (Oberley & Steinert, 1983; courtesy of Dr T. Oberley). A. Small, rounded cells in colonies when cultured in the presence of laminin. B. Flattened, dispersed cells subcultured with fibronectin.

(1983) included collagen digestion of minced pieces from guinea pig renal cortex, followed by sieving through a nylon screen and plating onto culture dishes. The chemically defined culture medium consisted of an ordinary salt solution supplemented with transferrin, insulin, selenium, and tri-iodothyronine. The tubular epithelial cells were explored for their adhesive and proliferative capacities in the presence of certain matrix molecules (Fig. 2.4).

Collecting-duct epithelial cells have been examined in primary cultures by a method developed by Minuth (1982, 1983; Minuth & Kriz, 1982). Fragments of the renal cortex of neonatal rabbits contain a fibrous capsule, S-shaped nephric bodies, and collecting-duct anlagen. *In vitro*, such fragments round up to 'globular bodies' and become surrounded by an epithelial sheet derived from the collecting-duct epithelium. While the other components degenerate, the collecting-duct cells develop an outgrowth of epithelial cells in a monolayer (Fig. 2.5). These cell cultures have been used for certain metabolic studies and for tests on the effects of various metabolic inhibitors (Minuth, 1983; Minuth *et al.*, 1984).

Successful cell cultures derived from bovine collecting-duct epithelium have been prepared by Gospodarowicz *et al.* (1983), and more recently by Perantoni *et al.* (1985), whose method produces good cell cultures from rat collecting-duct epithelium. Starting from the ureteric bud of 13- to

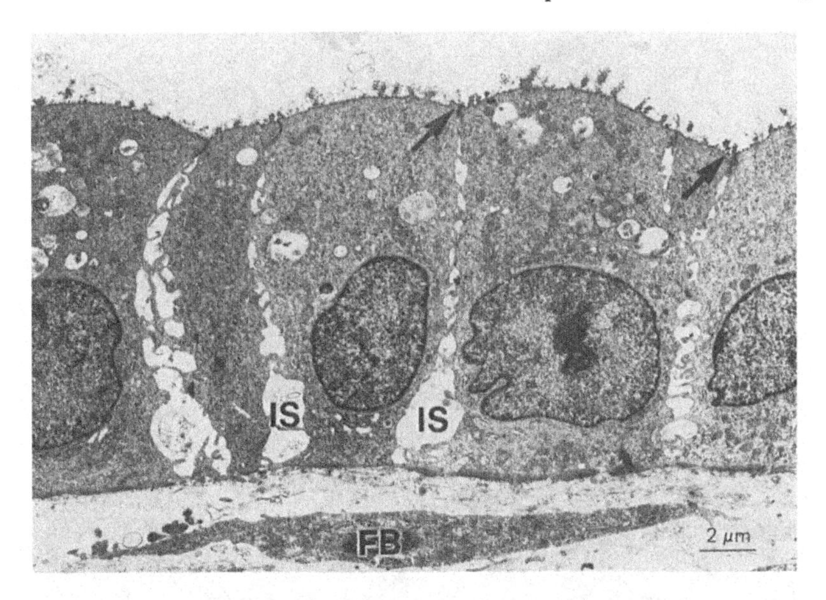

Fig. 2.5. Electron micrograph of polarized epithelial cells from the renal collecting duct epithelium (Minuth, 1983; courtesy of Dr W. Minuth). IS, intercellular spaces; FB, fibroblasts; arrowheads, tight junctions.

14-day rat embryos, these authors obtained cells expressing many ultrastructural and histochemical features that demonstrated a high degree of structural and functional maturity.

Differentiating mesenchymal cells from experimentally induced nephric blastemas can be brought into culture by a relatively simple procedure (Lehtonen *et al.*, 1985; Saxén & Lehtonen, 1986). Following the protocol described on p. 48, isolated metanephric mesenchymes are induced by the transfilter method for 24 h. When the induction is completed, the mesenchymes are cut into small fragments and transferred onto coverslips on the bottom of common culture dishes. No signs of epithelial transformation of the mesenchymal cells are observed at this stage, and the fibroblast-like cells are characterized by a vimentin-containing cytoskeleton that is typical of undifferentiated connective-tissue cells. The cells, however, follow the programme set during their induction, and in prolonged cultures they show epithelial features including synthesis of fibrillar cytokeratin. The cells also synthesize and deposit matrix proteins such as fibronectin and laminin (Fig. 2.6). Similar findings were obtained when cells from previously formed tubules were examined in monolayer outgrowths (Thesleff *et al.*, 1983).

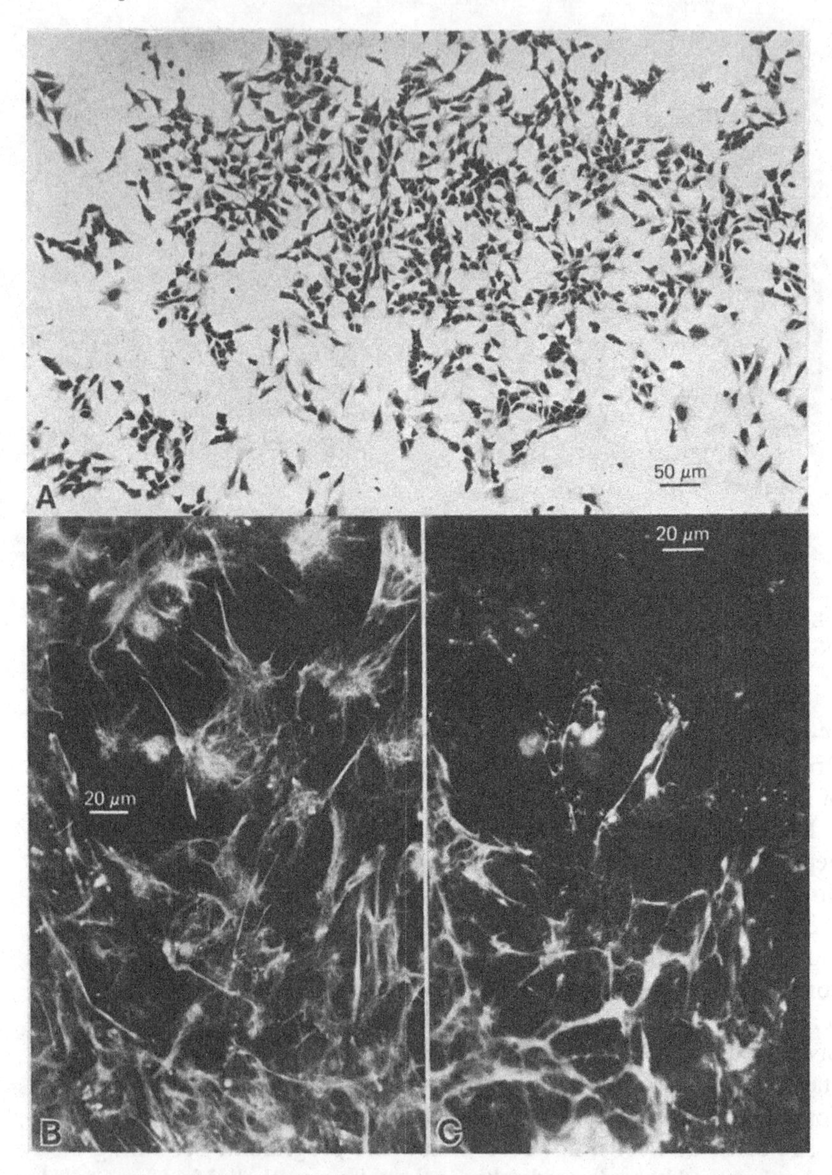

Fig. 2.6. Microphotographs of metanephric mesenchymal cells cultivated as monolayers for six days. A. Light micrograph of the culture stained with haematoxylin and eosin. B. High power fluorescence micrograph of mesenchymal cells treated with antiserum against vimentin. C. Fluorescence micrograph of the cells illustrated in B. after treatment with fluorochrome-conjugated antiserum against fibronectin.

Organ culture

While the above mentioned cell culture model-systems provide sensitive tools for an analysis of molecular and cellular events of cytodifferentiation, they have only limited value when aspects of three-dimensional cell assemblies during organogenesis are evaluated and when processes such as cell polarization and intercellular communication are to be explored. The organ culture technology was developed for such studies. In 1922, Rienhoff published his pioneering studies on chick nephric rudiments *in vitro*. He cut small fragments from the mesonephric and metanephric regions of 8- to 10-day chick embryos and cultured them in a hanging drop containing '85 cc Locke's solution plus 51 cc chicken bouillon plus 0.5 % dextrose'. Judging from his descriptions and illustrations of the cultures, both survival and growth were good within the explant as well as in the outgrowth of various cell types at its margins. Since then, many authors have cultivated avian nephric tissues, including the mesonephric anlage (e.g. Lash, 1963; Strudel & Pinot, 1965).

In 1950 Borghese published his excellent results on organ cultures of mouse salivary gland, and much of his experience was soon applied by Grobstein (1953*a*, *c*, 1955*a*), who analysed the mouse metanephros *in vitro*. Originally the dissected metanephric rudiments of 11-day mouse embryos were cultivated in Carrel flasks, which were soon replaced with flat culture dishes with a central well, the clot consisting of equal amounts of owl plasma and culture medium. The original medium consisted of Tyrode's solution, horse serum, and 9-day chick embryo extract in the proportions 2: 2:1. The medium was changed daily, and the dishes were kept in an atmosphere of 5% (v/v) CO_2 in air. In these conditions morphogenesis and good survival were obtained (Fig. 2.7).

We adopted these culture conditions and modified them slightly (Saxén *et al.*, 1962, 1968; Fig. 2.9). The dissected rudiments are placed on a piece of Millipore filter, which in turn is transferred to a Trowell-type screen of stainless steel. The screen is placed in a simple culture dish into which the culture medium is added up to the level of the screen. Due to surface tension, the tissue becomes covered with a thin layer of medium. The medium consists of Eagle's basal medium in Earle's balanced salt solution, supplemented with 10% (v/v) foetal calf (or horse) serum. The dishes are kept in an incubator with an atmosphere of 5% (v/v) CO_2 in air. The 11-day kidney rudiments survive well in these conditions up to 12 days and show good morphogenesis and advanced tubular structures (Fig. 2.8).

The different components of the culture medium are surveyed later. The best non-defined, protein-containing medium proved to be one that contained minimal essential medium (MEM) or Biggers medium (Biggers *et al.*, 1961) plus 10% (v/v) horse serum and 10% (v/v) chick embryo

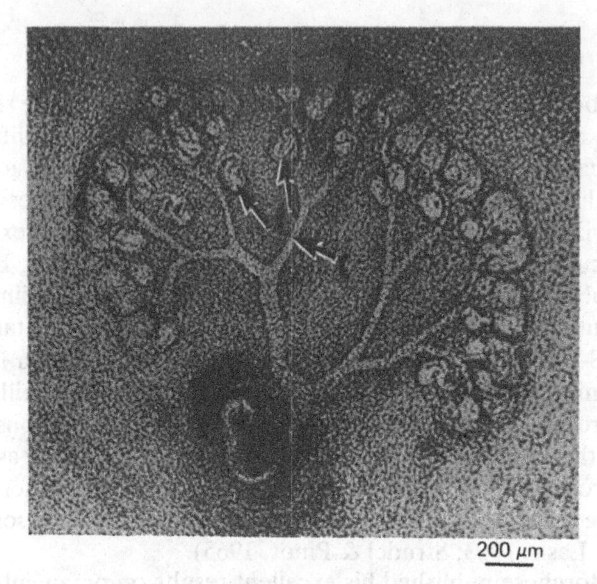

Fig. 2.7. Micrograph of an 11-day mouse metanephric rudiment after eight days of cultivation (Grobstein, 1955*a*). Upper arrows, S-shaped bodies; lower arrow, collecting duct.

extract (Saxén, 1983). A chemically defined medium has been used that consists of an enriched MEM (Richter *et al.*, 1972) supplemented with human transferrin (Ekblom *et al.*, 1981*c*, 1983). Differentiation in this medium was not quite as good as that in the serum-containing medium. A more complex defined medium for mouse kidney tissue has been devised by Avner *et al.* (1982). In addition to a commercial salt solution, the medium consisted of insulin, hydrocortisone, prostaglandin E, and transferrin.

Organ culture techniques have also been successfully applied to human embryonic tissues, including kidney. Large numbers of such experiments have been published, and various exogenous agents have been tested in this system, possibly providing methods for testing toxicity and teratogenicity (Crocker & Vernier, 1970; Lash & Saxén, 1972; Crocker, 1973; for a review, see Saxén, 1983).

For various types of experiments, especially those dealing with cell and tissue interactions, it is necessary to separate the components, the different cell lineages, within a tissue or organ anlage. In the case of the kidney, the separation is done between the mesenchymal blastema and the invading, still unbranched ureteric bud. For this, Grobstein (1955*a*) used a short-term treatment with 3% (v/v) trypsin followed by mechanical treatment consisting of repeated pipettings and gentle teasing with a cataract knife. Instead of the enzyme treatment, we use a Ca^{2+}- and Mg^{2+}-free saline containing 0.02% (v/v) EDTA (Versene), and the

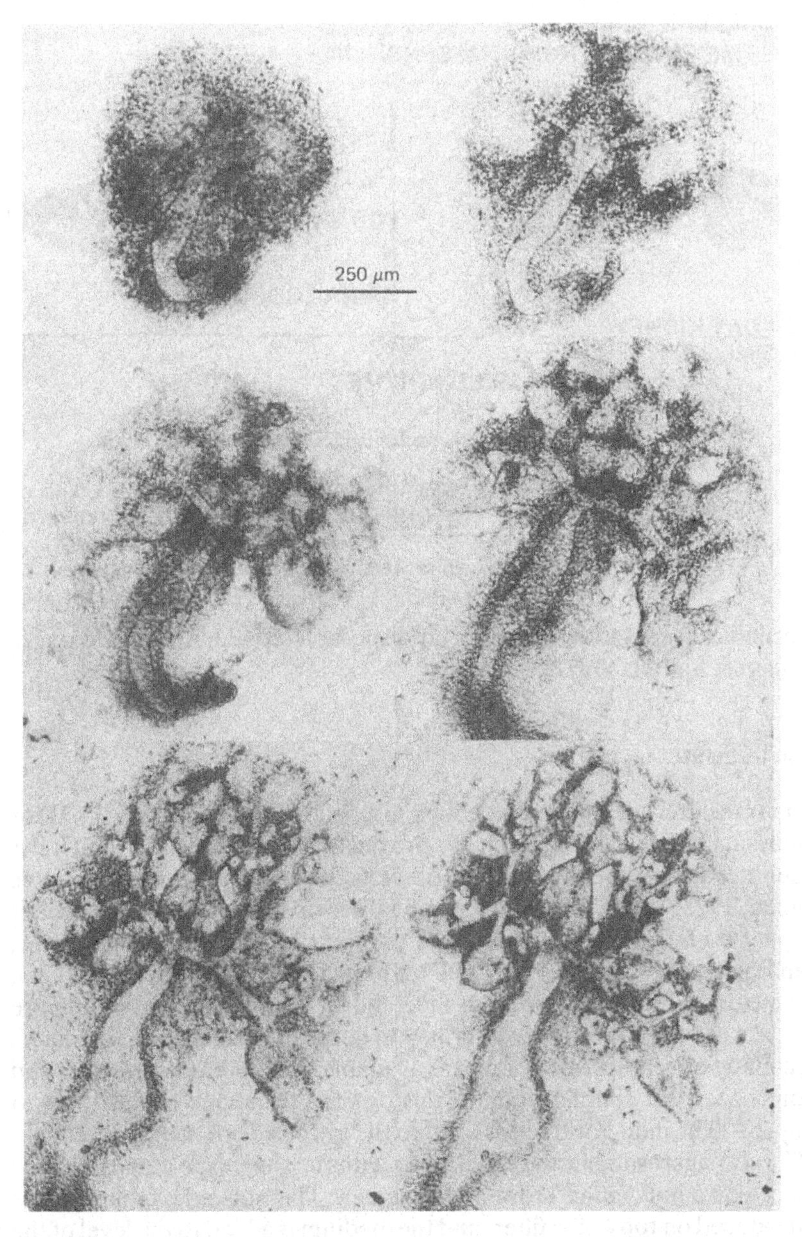

Fig. 2.8. Daily prints of a time-lapse motion picture of an 11-day metanephric rudiment cultivated *in vitro* (Saxén & Wartiovaara, 1966).

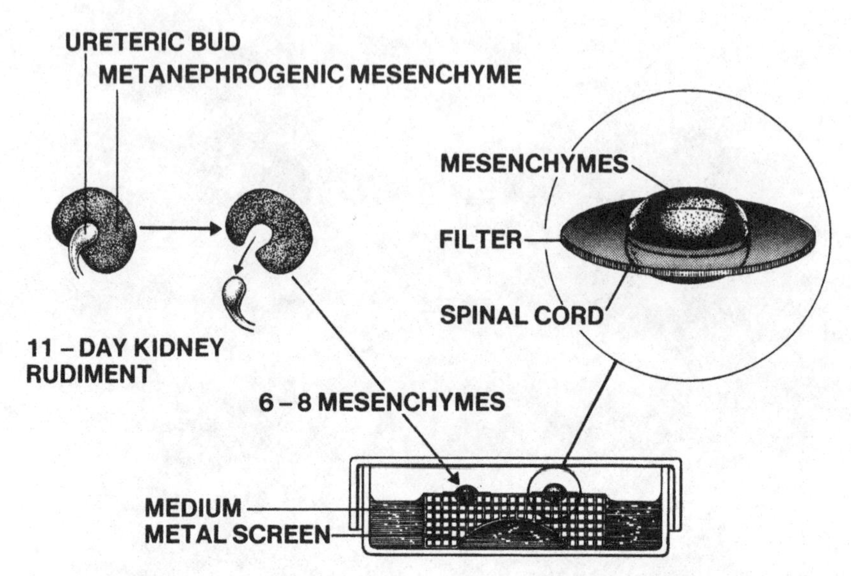

Fig. 2.9. Scheme of the transfilter technique described in the text.

mechanical separation is done with disposable no. 23 injection syringes (Saxén & Saksela, 1971).

The transfilter technique

The transfilter technique created by Grobstein (1953*b*, 1956*a*, 1957) for analysing kidney development has proved to be most effective, and it has been used for a great variety of other tissues as well (for reviews, see Saxén, 1972, 1980). Grobstein's original set-up consisted of the test filter supported by a plexiglass ring fixed to a pair of glass rods. The tissues, separated as described, were glued onto both sides of the filter membrane by a clot and cultivated as described above. Various types of Millipore filters were used, as will be described in detail in Chapter 3. These filters are made of cellulose ester and are available in different thicknesses and pore sizes. Our simplification of this original technique is illustrated in Fig. 2.9. The inductor tissue is fixed to the lower side of the test filter with 1% (v/v) agarose, and the filter piece without any supporting devices is placed on a hole in the Trowell-type screen. The isolated mesenchyme is introduced on top of the filter, and the medium is added to the level of the filter. The Millipore filters with a spongy structure and irregular pores were later replaced by commercially available Nuclepore filters (Wartiovaara *et al.*, 1972, 1974). These filters are made of polycarbonate tape exposed to charged particles in a nuclear reactor followed by chemical

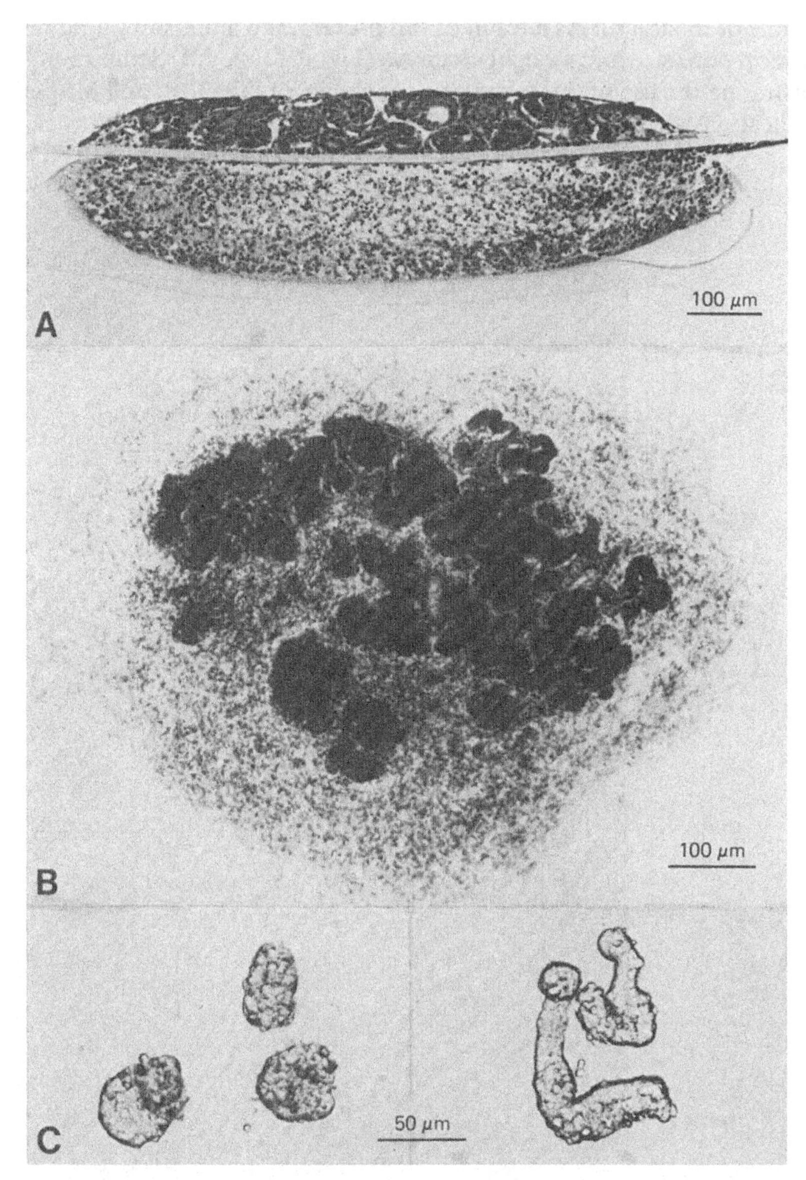

Fig. 2.10. Results of a transfilter induction of the metanephric mesenchyme by embryonic spinal cord. A. Section of an explant cultivated for 72 h. B. Similar explant viewed from top as a whole-mount preparation. C. Primitive tubules and renal vesicles enzymically separated from a mesenchyme induced for 72 h (Saxén & Wartiovaara, 1966).

etching of the tracks. As a result of this process, the filters show a rather uniform pore size and straight channels (Fig. 3.13, p. 76). Some results illustrating the use of this technique are shown in Fig. 2.10, and further applications are discussed in the following chapter.

3

Morphogenetic tissue interactions in the kidney

Introduction

In Chapter 1, I described the development of the two major components of the metanephric kidney, the branching epithelium of the ureter and the mesenchyme converted into epithelial elements. The development of these two cell lineages occurs in a strictly controlled, temporally and spatially synchronous manner. Theoretically, such control could be achieved in two ways; either both tissue components are under the same organismal control system, ensuring the timing of development, or the two cell lineages become aware of each other by exchanging signals that co-ordinate their development. I believe that the two mechanisms are interacting and are both involved in organogenesis, as will be discussed in Chapter 4. Here I will focus on the local interaction of the epithelial and mesenchymal components of the kidney.

General background

A morphogenetically significant interaction between two tissue components was demonstrated by Spemann (1901, 1912), who transplanted an optic vesicle onto a heterotopic site in an amphibian embryo, thus stimulating the formation of an extra lens from the overlying epidermis. This 'induction' could be prevented by interposing some impermeable material between the optic vesicle and the epidermis. Subsequently the concept of 'embryonic induction' was developed, and it became established in the 1920s after the demonstration of the primary inductive events during gastrulation (for reviews, see Spemann, 1936; Saxén & Toivonen, 1962). We know now that the inductive system operates throughout embryogenesis, 'whenever two or more tissues of different history and properties become intimately associated and alteration of the developmental course of the interactants results' (Grobstein, 1956*b*). Such morphogenetic interactions have been demonstrated as early as the four- to eight-cell stages of embryogenesis, and also in adult organisms during regenerative processes (for reviews, see Grobstein, 1967;

Fleishmajer & Billingham, 1968; Saxén & Kohonen, 1969; Kratochwil, 1972, 1983; Saxén *et al.*, 1976*a*, 1980; Wessells, 1977).

When morphogenetic interactions are dealt with in this volume and when their biological and molecular aspects are discussed, the presentation is largely restricted to the kidney and to the inductive interactions during nephrogenesis. It should be emphasized that the kidney is not necessarily the best model-system to study inductive processes, and no generalizations should be made on the basis of the data presented. There are few if any recognized similarities between the different morphogenetic interactions during development, and apparently a great variety of processes have been lumped together under the term 'induction'. As I have stated earlier, a search for a common determinative factor, or 'organizer', failed in the 1930s, and today we may fall into the same trap in searching for mechanisms common to the various types of inductive interactions (Saxén & Karkinen-Jääskeläinen, 1975). We should also keep in mind that, whenever we explore an inductive event in a simplified model-system, we are dealing only with a single step in a complicated chain of developmental processes leading ultimately to the formation of a complex, functional tissue.

Early observations of induction in the kidney

I have already presented some of the classic observations indicating that inductive tissue interactions are involved in early kidney development. An interaction between the cells of the pronephric duct and those of the surrounding tissues guide the formation and migration of the duct, as first suggested by Holtfreter (1944) and recently demonstrated by Poole & Steinberg (1982) (p. 7). Evidence on the interaction between the Wolffian duct epithelium and the mesonephrogenic mesenchyme is somewhat inconsistent, but it strongly suggests that such an interaction is important for the completion of mesonephric development. It might be necessary, however, to emphasize that all experimental results on this interdependence do not fully agree. In amphibians especially, varying observations have been reported (Table 3.1). According to Humphrey (1928), grafts of urodele mesonephric blastema develop tubules in complete absence of the Wolffian duct, but other workers have shown that nephrogenesis comes to a standstill after elimination of the duct (Miura, 1930). So, it has been suggested that tissues other than the Wolffian duct may also act as inductors during early development of the mesonephros (Torrey, 1965). In fact, there is evidence from both amphibians and birds that endoderm may act as an inductor upon the mesonephric mesenchyme (Etheridge, 1968; Croisille *et al.*, 1976).

The interdependence of the two components of the metanephric blastema can explain the development of some congenital defects of the

Table 3.1. *Summary of experimental results on the role of the nephric duct in the formation of mesonephric tubules*

Experiment	Form used	State of mesonephric tubules	Reference
Grafts of mesonephros – no duct present	*Amblystoma*	Well-differentiated tubules	Humphrey, 1928
Excision of nephric duct	*Rana*, *Bufo*	No development of tubules	Miura, 1930
Excision of nephric duct	*Bufo*	Irregular differentiation of tubules	Shimasaki, 1930
Excision of pronephros and duct	*Amblystoma*	Irregular differentiation of tubules	Burns, 1938
Excision of pronephros and duct	*Rana*	Local condensations of cells	Waddington, 1938
Obstruction of duct primordium	*Rana*, *Pleurodeles*	Local condensations of cells No condensations present	O'Connor, 1939
Obstruction of growing duct	*Triton*, *Amblystoma*	Local condensations of cells	Holtfreter, 1944
Excision or obstruction of duct	*Rana*	Cellular blastema forms – disappears	van Geertruyden, 1946
Removal of duct after formation	*Rana*	Local condensations of cells	
Partial removal or displacement of duct	*Alytes*	Tubules develop only in close proximity to duct	Cambar, 1948
Excision of nephric duct	*Triturus*	No development of tubules	Kotani, 1962
Transplantation of mesonephric mesenchyme devoid of nephric duct	*Triturus*	Frequent formation of tubules	Etheridge, 1968
Tip of duct destroyed by cautery	Chick	Local condensations of cells which later disappear	Boyden, 1927, 1932
Tip of duct obstructed by incision	Chick	Local condensations of cells	Waddington, 1938
Tip of duct destroyed by cautery	Chick	No tubules caudal to lesion	Gruenwald, 1937
Tip of duct blocked by graft	Chick	Irregular differentiation of tubules	Gruenwald, 1942
Coelomic grafts of caudal fragments devoid of nephric duct	Chick	Tubule formation in 43%	Croisille *et al.*, 1976

After Burns, 1955, completed.

urinary tract (Auer, 1947; Gluecksohn-Schoenheimer, 1949; Wharton, 1949). The ultimate experimental proof for this inductive interaction between the ureteric epithelium and the metanephrogenic mesenchyme came from Grobstein, who in 1953 reported the first results of separation and recombination experiments, as described on p. 48. Isolated kidney epithelium did not show its characteristic morphogenesis when combined with submandibular capsular mesenchyme, but did so when re-combined with its own metanephric mesenchyme. In the latter set of experiments, 'coiled tubules' were detected in the mesenchyme around the branching ducts. These pilot studies were repeated and amplified, and, in a series of classic papers, Grobstein outlined the basic features of kidney tubule induction (Grobstein, 1953*c*, 1955*a*, *b*, 1956*a*, *b*, 1957, 1961).

The mesenchyme

In the vertebrate embryo, the nephric mesenchyme forms a paired ridge (Fig. 1.3, p. 3), the most caudal part of which will contribute to the metanephric anlage. As already shown, this mesenchymal component differentiates into the secretory part of the nephron. It also affects the epithelial component morphogenetically, supporting the branching and proliferation of the latter.

Effect on the epithelium

When cultivated *in vitro*, a separated epithelium soon flattens and spreads without signs of branching or morphogenesis (Grobstein, 1955*a*). When, on the other hand, the separated epithelium is cultivated in association with a cluster of trypsin-separated metanephric mesenchymes, branching is initiated. In avian embryos a similar but incomplete ramification of the Wolffian duct was reported when it was combined with the mesonephric mesenchyme (Martin, 1976), but no heterologous mesenchymes have been reported to support the epithelial morphogenesis.

The effect of kidney mesenchyme on the morphogenesis of the epithelium is an example of an epithelial–mesenchymal interaction demonstrated in practically every glandular and many other tissues containing the two components. In the kidney, the mesenchymal influence seems to be rather specific, whereas this specificity varies in different organ anlagen. In the salivary gland, the effect still shows relative specificity (Grobstein, 1967; Lawson, 1972), but as Lawson (1974) later showed, a heterologous membrane (lung) also supported branching of the salivary epithelium if the mass of the pulmonary mesenchyme was sufficiently large. Lawson suggests that the varying results are due to the mass of the mesenchyme used, culture conditions, and the fact that no mesenchyme-specific factors are involved.

The pancreas is an extreme example of non-specificity of the mesenchymal contribution to epithelial morphogenesis. After the 30-somite stage in the chick, the pancreatic epithelium still requires exogenous support, which is normally provided by the homologous mesenchyme. This can be replaced experimentally by almost any heterologous mesenchyme, by high concentrations of chick embryo extract or by a protein preparation promoting cell proliferation (Rutter *et al.*, 1964, 1967; Fell & Grobstein, 1968; Pictet *et al.*, 1975).

In order to explain the varying specificity of the mesenchymal inductive action, Grobstein (1967) postulated that two active factors were involved, the mesenchyme-common factor and the mesenchyme-specific factor, and that the latter 'may contain developmentally significant information'. Grobstein's hypothesis might still be considered, as both types of influence have been shown to exist. The mesenchyme-specific action is apparently found only in the homologous mesenchyme, but additional factors (mesenchyme-common?) are required *in vitro*: when a protein-free medium is used, branching is incomplete or totally blocked, but addition of serum or transferrin leads to active branching (Thesleff & Ekblom, 1984).

In the metanephric kidney, the mesenchymal action on the epithelium has not yet been examined in detail, but perhaps an analogy may be found in another model-system, the embryonic salivary gland, where a similar, epithelial branching occurs. Here, the mesenchyme acts on the epithelial basement membrane causing its degradation and remodelling at the branching sites (Bernfield *et al.*, 1984*a*, *b*). However, separation and recombination experiments of various types suggest that the primary 'branching programme' is in the epithelium itself.

Competence and developmental options

Interactive events in the kidney model-system are only links in a long chain of developmental processes, and so the prehistory of the mesenchyme and its possible previous commitments should be explored. Is the mesenchyme still uncommitted, with many developmental options, or have these options become restricted during early development?

Hematopoiesis and granulopoiesis have been detected in the amphibian pronephric anlage. The former is due to external colonization of the pronephros, but the granulopoiesis has been suggested to be an intrinsic property of pronephric tissue (Carpenter & Turpen, 1979; Turpen & Knudson, 1982). In the mouse metanephric mesenchyme, a hematopoietic and endothelial bias has likewise been described (Emura & Tanaka, 1972), but we have not been able to confirm these results

(p. 131). A chondrogenic potency has been demonstrated in avian and human mesonephric mesenchyme (Lash, 1963; Lash & Saxén, 1972) but not in the metanephros. Thus, to the best of my knowledge, the metanephric mesenchyme has developmental options definitely restricted either to remaining a mesenchymal stroma or to being converted into epithelial tubules.

If the view is accepted that the metanephrogenic mesenchyme, prior to contact with the ureteric epithelium, has become 'predetermined' and possesses a nephrogenic bias without other developmental options, it should be asked when and how this early commitment has occurred. This inevitably brings us back to the early stages of nephrogenesis and to the development of the pronephros and the mesonephric blastema described in Chapter 1. Two types of experiments have been performed with pronephric and mesonephric anlage tissue to answer the above question: early stages of the nephric blastema are either totally dissected and cultivated *in vitro* or *in vivo* or they are exposed to various heterologous tissues or grafting conditions.

The presumptive pronephric area has been mapped out in early amphibian gastrulae by a vital stain method (Fig. 1.3, p. 3). This, of course, does not necessarily imply that the anlage area has already been determined at this early stage. If the blastoporal lip area containing the prospective pronephric territory is dissected out, disaggregated and subcultivated in reaggregated form, nephric tubules indeed develop (Holtfreter, 1944), but they may have been determined by interactions between various tissues during prolonged subcultivation of the reaggregate that contains both neural and mesodermal structures. Similarly, if the prospective, multipotent gastrula neuroectoderm is exposed to various heterogeneous inductors *in vitro*, nephric tubules often form, as shown in Fig. 3.1. However, only in rare instances does tubule formation become the only consequence of the induction, and, as a rule, other structures develop as well (Toivonen & Saxén, 1955; Saxén & Toivonen, 1961). Hence, it is feasible that the tubules are formed again as a consequence of a secondary induction rather than having been converted directly from the competent ectoderm.

The nephrogenic potency is not restricted to the still uncommitted ectoderm (above). In fact, almost any part of the ectoderm, the endoderm, and the mesoderm of young amphibian gastrula will form nephric tubules when transplanted to the prospective region, the 'nephric field' of an embryo (Holtfreter, 1933).

In chick embryos, the pronephric mesenchyme might be determined as early as the early neural plate stage, when the intermediate mesenchyme can form pronephric tubules in the absence of the somitic and the lateral plate mesoderm (Waddington, 1938). These dissection experiments cannot, however, rule out fully the role of neural tissue in the induction of

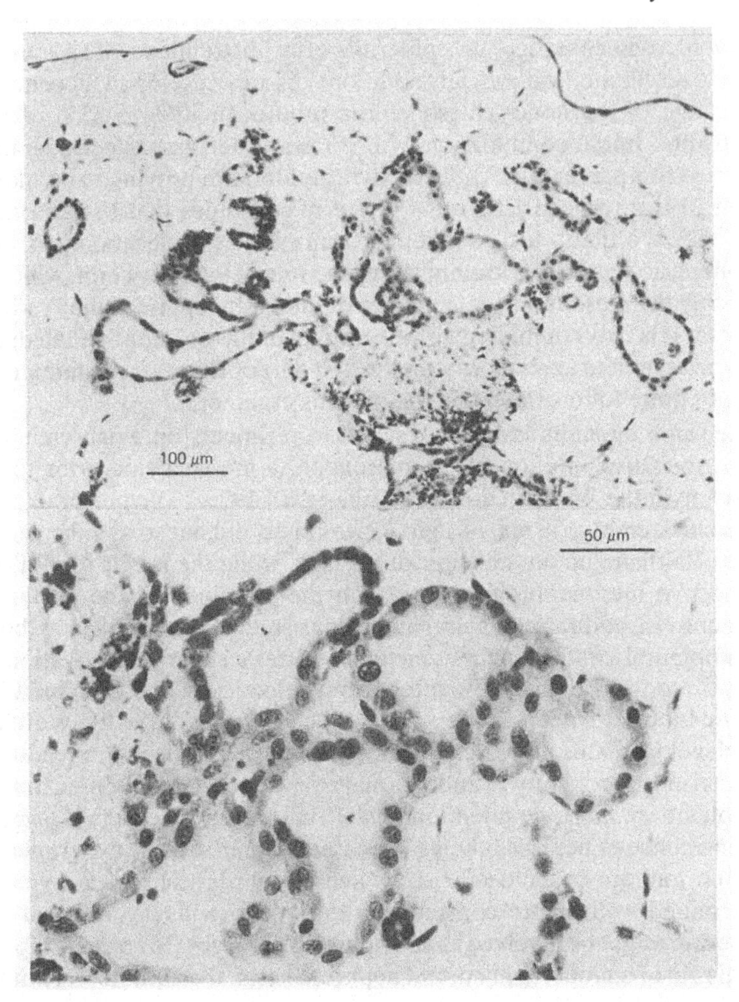

Fig. 3.1. Micrographs showing abundant kidney tubules in an explant made by combining a piece of prospective gastrula neuroectoderm with a devitalized heterologous inductor, a fragment of guinea pig bone-marrow (Toivonen & Saxén, 1955, and unpublished results).

nephric tubules, as stated by Waddington himself. The prospective mesonephric area can likewise be traced back to the gastrula stage of amphibian embryos (Fig. 1.3, p. 3), but mesonephric tubules can be obtained from other sources also, as shown by Spofford (1948). The stage of the first determination of the mesonephric anlage is not fully explored, nor do we know how this is implemented. To answer these questions, Etheridge (1968) performed extensive experiments with newt embryos from young gastrulae to neurula stages. The prospective mesonephric mesenchyme was transplanted either alone or combined with other embryonic tissues under the epidermis of a host embryo. The mesenchyme transplanted without any additional tissues developed, depending on the age of the donor, mesonephric tubules in 17% to 55% of the transplants. Initial combination with certain other tissues considerably increased this percentage, the underlying endoderm proving particularly potent in this respect. Etheridge properly concludes that tissues other than the Wolffian duct epithelium can act as inducers upon the mesonephric blastema. He emphasizes the role of the endoderm, which is 'probably the most important inductor of the mesonephric kidney', while the process is only completed by the terminal inductor, the Wolffian duct. While in complete agreement with the first part of Etheridge's statement, I cannot quite follow the second part of his argument.

The same dilemma is met in related experiments on avian embryos involving transplantation of the mesonephric mesenchyme prior to its contact with the Wolffian duct (Croisille *et al.*, 1976). The posterior part of the chick embryo at stages 15 to 25 was dissected out so as to leave the entire Wolffian duct in the anterior portion, while the major part of the prospective mesonephrogenic mesenchyme remained in the posterior fragment. In addition, the fragment consisted of ectoderm and both somatopleural and somitic mesenchyme. After a short precultivation *in vitro*, these fragments were grafted intracoelomically and examined for nephric tubules. Unexpectedly, 43% of the grafts devoid of the Wolffian duct developed mesonephric tubules. Since these tubules were regularly found in close proximity to endodermal derivatives, Croisille *et al.* tested the role of the latter by intentionally leaving the endoderm in the graft. The incidence of nephric tubules increased further in these experiments, and the authors concluded that the mesonephrogenic mesenchyme is 'determined well before contact with the Wolffian duct', and that the endoderm might be involved as in amphibian embryos. In considering the actual role of endoderm in normal nephrogenesis, the authors were more cautious.

The early determinative steps of the metanephrogenic mesenchyme are not known, but there are some analogies with the mesonephric blastema. As will be discussed in more detail below, the metanephrogenic mesenchyme, prior to its contact with the ureteric bud, is already predetermined, restricted in its developmental options, and responsive to various heterologous inductors. But it is not known when and how this determination has occurred.

Homogeneity

Despite the restricted options of the mesenchyme, the epithelialized mesenchymal cells have retained options for continued differentiation. *In vivo*, they give rise to many special types of cells within the nephron and, in prolonged cultures *in vitro*, at least three distinct epithelial phenotypes have been characterized (Ekblom *et al.*, 1981*a*). These are the podocytes of the glomeruli and the cells of the proximal and distal tubules. Before induction of these various cell types can be explored, it should be known whether their programmed precursors are present in the non-induced metanephric mesenchyme, or whether the latter is a homogeneous population of cells with identical developmental options.

These questions have not yet been answered conclusively, but they will be discussed later. Many observations suggest that the mesenchyme is uniform as far as the status of its cells is concerned: light and electron microscopy and immunohistology with probes for several matrix and cytoskeletal proteins have revealed only one cell type in the uninduced mesenchyme (Wartiovaara, 1966*a*; Ekblom, 1981*b*; Lehtonen *et al.*, 1985). Furthermore, cell electrophoresis has failed to show cell populations with varying surface-charge densities (Saxén *et al.*, 1965*b*), and monolayer outgrowths of mesenchymal explants show only one cell type when examined for the synthesis of some matrix proteins and cytoskeletal components (Lehtonen *et al.*, 1985, unpublished results). This observed homogeneity of the mesenchymal cell population cannot, however, exclude the possibility of subpopulations with varying developmental programmes already set.

The inductor

After concluding that the metanephric mesenchyme consists of 'predetermined' cells with an epithelial bias, the triggering inductor should be explored. We will examine whether the inductive capacity is restricted to the normal inductor, the ureteric bud, or whether heterologous tissues can mimic the effect (as do many tissues and their subfractions, in primary induction). We may also ask whether the response of the mesenchyme is of an 'all-or-none' type, i.e. quantitatively and qualitatively independent of the type of the inductor tissue.

Specificity

The original observations reviewed above and the experimental results of Grobstein (1953*c*) showed conclusively that the normal inductor of the kidney tubules is the ureter, whereas the studies of Gruenwald (1943, 1952) had suggested that neural tissue might have a similar effect. Grobstein (1955*a*) confirmed this, and since then embryonic spinal cord has been used as an active inductor. Its inductive effect has been demonstrated also by combining the spinal cord with amphibian and avian mesonephric mesenchyme (van Geertruyden, 1946; Croisille *et al.*, 1976). In addition, a vast number of other tissues and their fractions have been tested, of which many have proved inductively active when combined with the metanephric target mesenchyme.

Table 3.2. *The inductive action of certain tissues tested in combination with the metanephric mesenchyme of 11-day mouse embryos*

Tissue	Active	Inactive
Embryonic epithelia		
Ureter	+	
Submandibular	+	
Pulmonary	−	
Gastric	−	
Pancreatic	−	
Neural tissues		
Embryonic spinal cord	+	
Embryonic medulla	+	
Embryonic brain	+	
Embryonic spinal ganglia	−	
Embryonic spinal cord[a]	+	
Adult brain	−	
Neural teratoma	+	
Embryonic mesenchymes		
Salivary	+	
Jaw	+	
Head	+	
Tail		−
Limb bud		−
Developing bone	+	
Embryonic and adult liver		−
Adult retina and iris		−
Adult kidney tubules		−

[a] Chick origin, other tissues are murine.
Data from Grobstein, 1955*a*; Unsworth & Grobstein, 1970; Lombard & Grobstein, 1969; Auerbach, 1972; Saxén *et al.*, unpublished data.

Perusal of the list of tissues with an inductive action (Table 3.2) gives no clues as to what features of the tissues might be associated with their inductive capacity. Embryonic tissues are apparently more potent than adult tissues (Lombard & Grobstein, 1969), and many show variation in their inductive action depending on their stage of development (Grobstein, 1955*a*; Unsworth & Grobstein, 1970), but it is not possible to draw any further conclusions. Many of the tissues that show inductive activity are never associated with the nephric blastema during normal development, and thus, their inductive activity is biologically meaningless. Apparently the 'tubulogenic bias' of the nephric mesenchyme can be triggered by factor(s) widely distributed in embryonic tissues and basically serving other purposes.

By definition, the response to a kidney inductor is the formation of epithelial tubules that mimics normal development *in vivo*. Many events at the cell and molecular levels are associated with this process (see

Chapter 4), and whether the response is similar when artificial inductors are used, or whether inductor-dependent variations of some type exist, should be examined.

Gruenwald (1943) reported that when he combined the metanephric mesenchyme with nervous tissue, the resulting tubules were of mesonephric type. As a possible explanation, he suggests that a strong induction might induce a rapid mesonephric type of development, whereas a weaker effect would lead to the formation of metanephric elements. This line of research has not been followed up, but there is a possibility that the final outcome of an induction is determined by the changing responsiveness of the target tissue and the kinetics of the induction. Lombard & Grobstein (1969) and Unsworth & Grobstein (1970) noticed that overt morphogenesis of the metanephric mesenchyme followed a different time-course after exposure to different inductors. Some heterologous inductors (e.g. brain) brought the mesenchymal differentiation to a stage of condensation only, without further tubule differentiation. The different time-course and response might be artificial and reflect metabolic changes during the cultivation of the inductor. The same tissue might also show changes in its inductive action during development.

Effect on heterologous mesenchymes

The inductor has a trigger-type action on a predetermined nephric mesenchyme with a tubule-forming bias. Considering this, one might not expect a similar response in non-nephric mesenchymes. Holtfreter (1944) and van Geertruyden (1946) demonstrated this negative response in amphibians, and therefore the results by Bishop-Calame (1965*a*, *b*) were unexpected. She associated chick ureter with various heterologous mesenchymes and grafted the combined explants on CAM. After such treatment the pulmonary mesenchyme formed 'coiled tubules and Malpighian glomerules', while the ureter branched and became multilayered. She reported a similar response in proventricular mesenchyme combined with the ureter.

The above results led inevitably to the conclusion that the inductors 'not only act as pace-makers, but promote specific differentiation factors' (Wolff, 1966). This is important for any further endeavours to characterize the inductor, and I therefore re-evaluated the results in the mouse metanepric model-system (Saxén, 1970*a*).

The undifferentiated mesenchymal component was dissected from embryonic lung, salivary gland, and ventricle and combined transfilter either to the ureteric bud from an 11-day metanephros or to a piece of spinal cord from the same embryos. None of the 63 cultures showed any tubular differentiation, and the mesenchyme developed into an

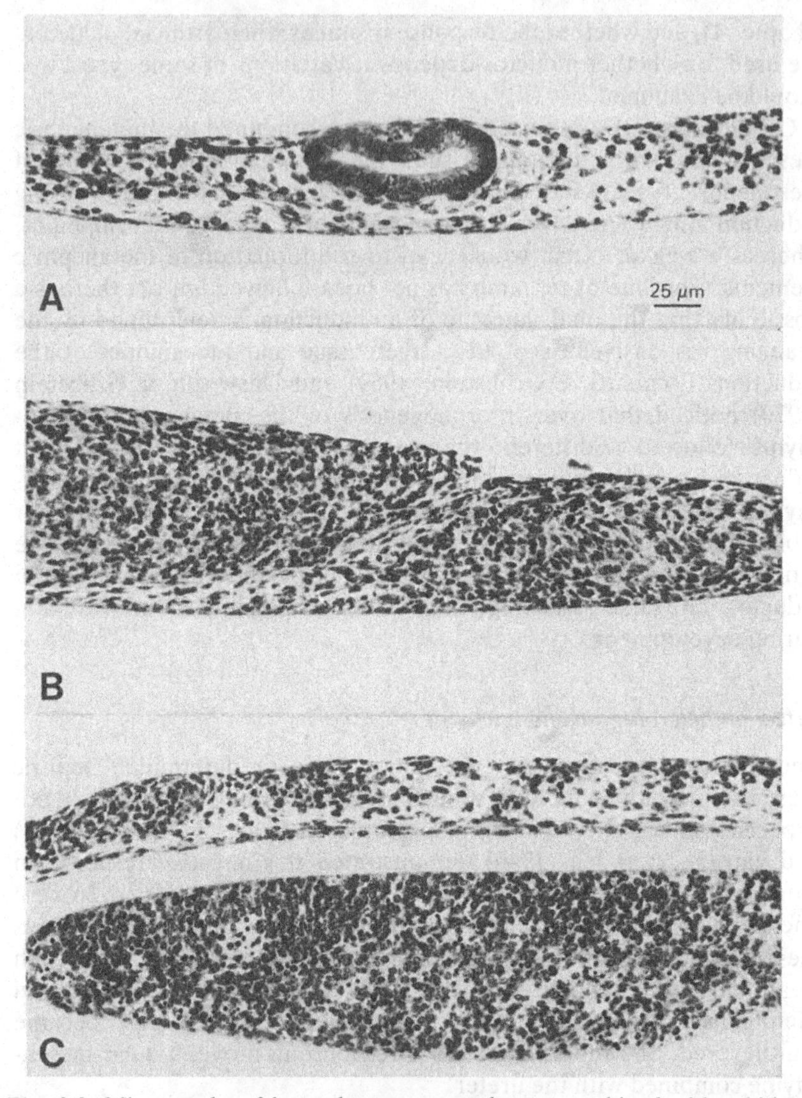

Fig. 3.2. Micrographs of heterologous mesenchymes combined with a kidney tubule inductor and cultivated for four days. No response can be detected in the mesenchyme (Saxén, 1970*a*). A. Isolated ureteric bud in combination with salivary gland mesenchyme. B. Direct combination of pulmonary mesenchyme (left) and spinal cord. C. Transfilter combination of salivary gland mesenchyme and spinal cord.

undifferentiated sheet of cells on the filter (Fig. 3.2C). The results were confirmed by combining mesenchymes from the same sources directly with spinal cord or with a ureter enzymically cleaned from mesenchymal cells. None of the 72 explants showed any tubules (Fig. 3.2A,B). When the experiment was repeated, using ureters mechanically separated but not cleaned by the enzyme treatment, tubules were found in 16 of the 36 cultures. The conclusion, therefore, was that the last result was due to contaminating mesenchymal cells on the ureter which were probably nourished by the heterologous mesenchyme.

In order to imitate the 'contamination' situation, we used an earlier, accidental finding in our transfilter experiments. If the mesenchyme is scraped from the filter after a 24-h exposure to the inductor, a number of cells in one or two layers usually remain attached to the filter, and occasionally they form small tubules. When these remaining cells were coated with a heterologous mesenchyme, tubule formation occurred regularly, and the tubules were more abundant and larger than in the uncoated cultures (Fig. 3.3). The possibility of a second-step induction by the nephric cells was excluded in two types of experiments. First, a double filter was used, and the contact between the two was broken, and the heterologous mesenchyme was placed on the bottom filter. No tubules were formed. The second experiment used chick cells, which are easily distinguishable from mouse cells. Chick tissue was placed on top of the mouse mesenchymal cells remaining on the filter. After two to three days of subcultivation, tubule-forming mouse cells were detected deep in the chick tissue (Fig. 3.4). The result gives further support to the transfilter results. Taken together, these experiments strongly suggest that the kidney tubule inductors (the ureter and the spinal cord) act exclusively on the metanephric (and mesonephric) mesenchyme, but cannot cause differentiation in heterologous mesenchymes.

Directive and permissive inductions

I have defined the two types of induction mentioned in the title (Saxén, 1977): 'When an embryonic cell possesses more than one developmental option, the choice between them is affected by extracellular factors, which thus exert a true *directive* action on differentiation. A *permissive* action, on the other hand, refers to a step of development, in which the cell has become committed to a certain pathway, but still requires an exogeneous stimulus to express its new phenotype' (Fig. 3.5). An example of the latter has already been given: pancreatic epithelial differentiation after a certain stage requires an extracellular stimulus, but this is highly unspecific and can no longer alter the course of differentiation (p. 55). The elegant studies of Cunha (1975) might be cited as an example of a directive influence.

Morphogenesis of the vaginal epithelium is guided by an epithelial–mesenchymal interaction, but at an early stage of differentiation the developmental options of the epithelium are not totally restricted. For example, its combination with uterine stroma leads to a uterus-type

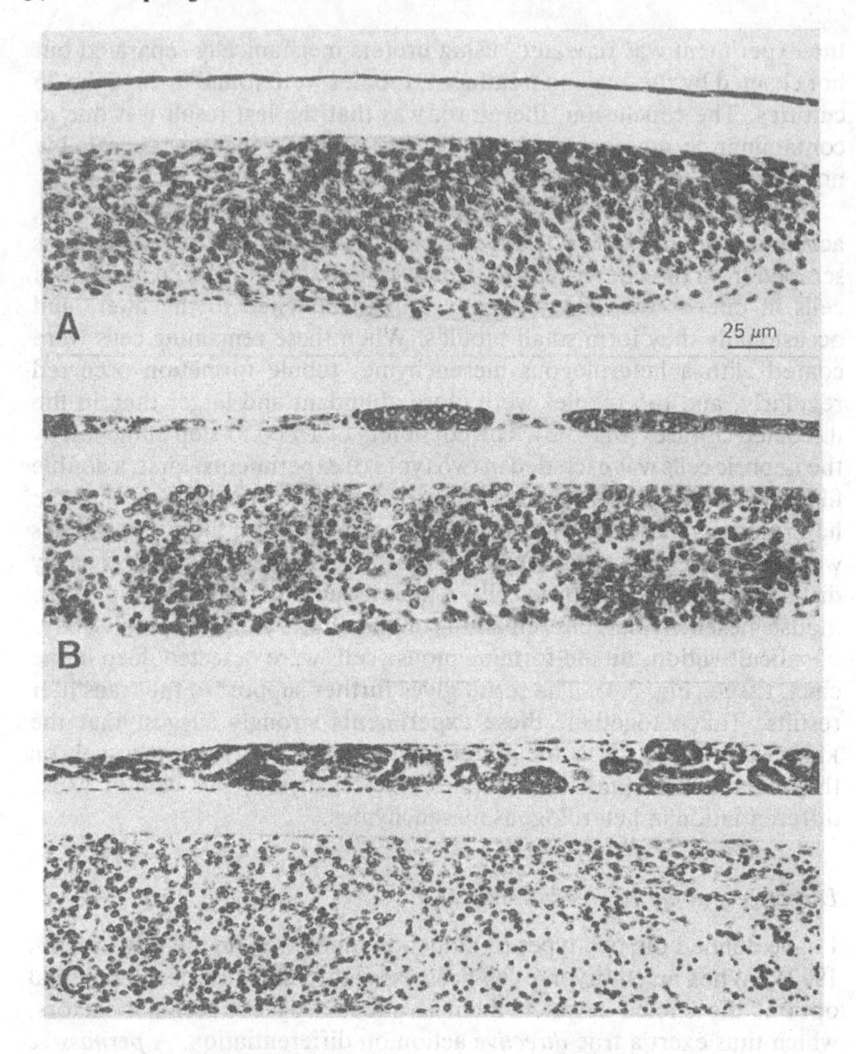

Fig. 3.3. Micrographs of cells remaining on a filter when the mesenchyme is removed after 24 h of transfilter culture. Subculture for three days (Saxén, 1970*a*). A. Thin sheet of undifferentiated mesenchyme cells remaining on the filter. B. Single small tubules occasionally developed in uncovered cultures. C. Abundant, well-differentiated tubules when the remnant cells were coated with a heterologous mesenchyme.

differentiation of the epithelium, and similar flexibility and directive actions have been demonstrated in the epithelium of the urinary bladder (Cunha, 1985).

My distinction between a directive and a permissive induction has been criticized by Kratochwil (1983). Though I may not agree with his views, I

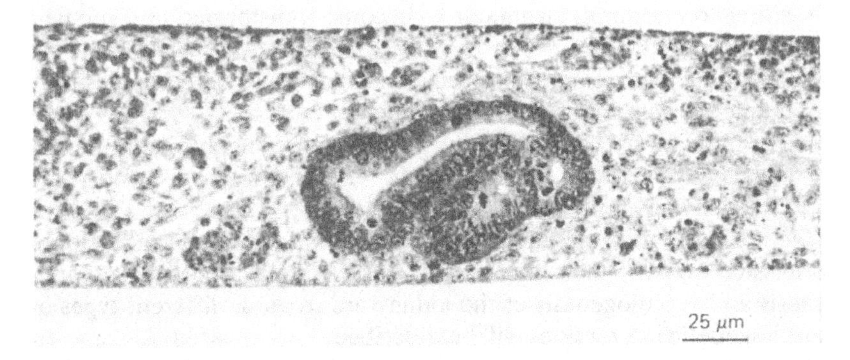

Fig. 3.4. Micrograph of an explant where the mouse mesenchymal cells on the filter were coated with chick mesenchymal cells distinguishable from mouse cells by their small, pale nuclei. Well-shaped mouse tubules develop (Saxén & Karkinen-Jääskeläinen, 1975).

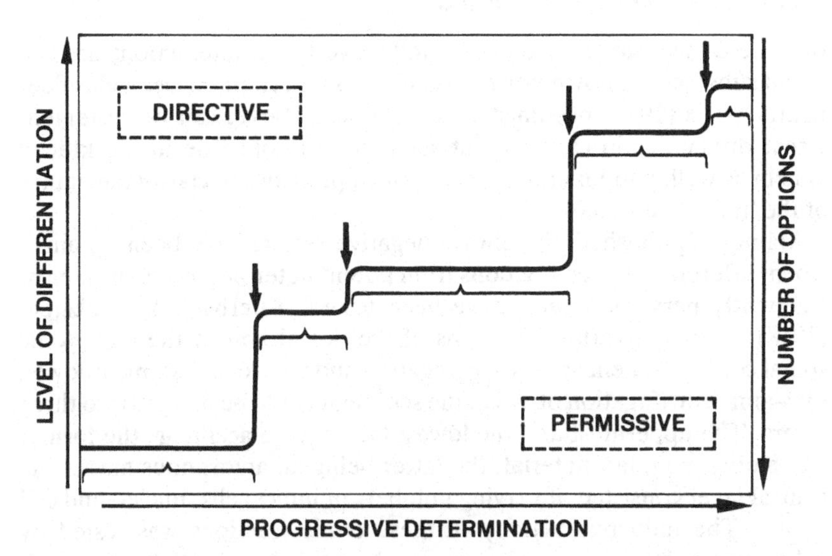

Fig. 3.5. Scheme of the hypothesis of 'directive' and 'permissive' inductions (after Saxén, 1977, 1981).

fully agree with his closing comment: 'In the absence of any molecular understanding of the process of developmental decision, and of the basis of developmental options or determination, our thinking about induction must be based on purely phenomenological information and it essentially relies on two testable properties of inductive systems: (1) the competence of the responding tissue and (2) the requirement of inductor specificity.'

According to these suggestions, the following conclusions on the kidney tubule induction may be summarized.

(1) The nephrogenic mesenchyme has only two developmental options: either to remain as stroma or to become transformed into tubule-forming epithelium.
(2) The inductive action is unspecific, as many non-related tissues fully mimic the action of the normal inductor.
(3) The inductor acts only upon the predetermined nephrogenic mesenchyme and exerts no detectable effects on other embryonic mesenchymes.

On the basis of these findings, I consider kidney tubule induction a permissive event. When later in this book certain aspects of organogenesis and vasculogenesis of the kidney are treated, different types of morphogenetic interactions will be described.

Inductive signals

Type and distribution of the signals

In none of the model-systems for inductive tissue interactions are the signal substances known nor has their action mechanism been clarified. Kratochwil's (1983) pessimistic view, quoted above, can be sustained. Great difficulties in carrying out such studies hold true in the kidney system as well, and I mention only some approaches to clarify the nature of the 'inductive signals'.

Mainly unpublished and always negative results have been obtained when different cell-free fractions from potent heterologous tissues, most frequently nervous tissues, have been tested. Auerbach & Grobstein (1958) examined various fractions of the dorsal half of the embryonic spinal cord. After enzymic disaggregation and vigorous shaking followed by 8-min centrifugation of 375 *g*, the sediment could be divided into three layers. The uppermost and the lowest layers were acellular, the former containing granular material, the latter being an amorphous mass. The central layer consisted of varying numbers of intact cells, nuclei, and cell debris. The inductive activity of these crude fractions was tested by cultivating isolated metanephric mesenchymes in hanging drops that each contained one of the three preparations. The supernatant and the two cell-free fractions of the sediment gave invariably negative results, while tubules developed in cultures containing the middle-layer material. Interestingly, the cellular fraction exerted an inductive action only as long as it contained viable cells, but became inactive when these disappeared.

Various 'conditioned' media have been tested, but only one, still-unconfirmed, positive observation has been reported. Auerbach (1977) exposed isolated metanephric mesenchymes to the supernatant from cultures of neural teratoma cells and obtained tubule formation. Unfortunately, this important finding has not yet been published as an

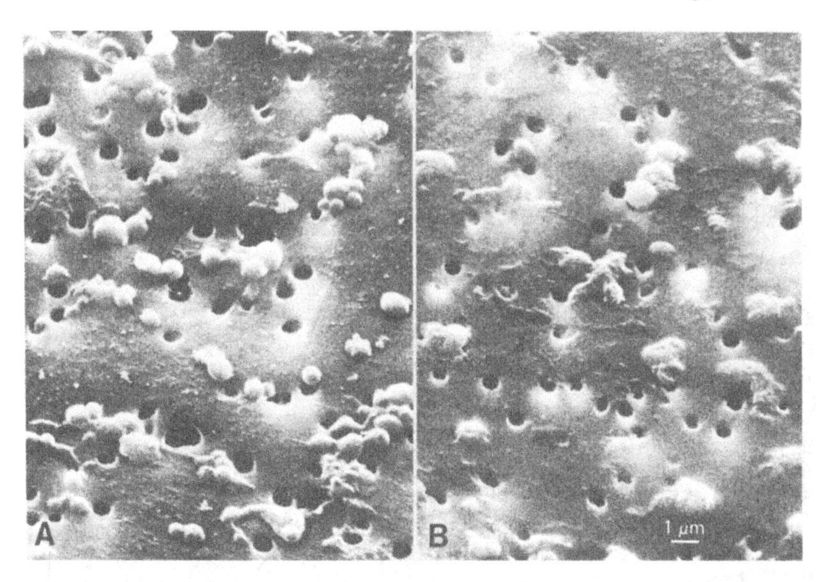

Fig. 3.6. Scanning electron micrographs of the upper surface of a Nuclepore filter with an average pore diameter of 0.5 μm cultivated for 24 h with a piece of spinal cord attached to the lower surface (Saxén & Lehtonen, 1978). A. Abundant cytoplasmic processes bulging through the pores. B. After an additional 24-h subculture in the absence of spinal cord, material is still left on the filter.

original article, and some details are missing from the available congress report.

We made another approach to test cell-free material by using the filter technique (Saxén & Lehtonen, 1978): when a piece from the dorsal spinal cord is cultivated for 24 h on (or under) a filter with a proper pore diameter, cytoplasmic processes penetrate the filter and bulge on the other side (Fig. 3.6A). After careful removal of the spinal cord tissue, these processes and their membrane material are left behind and will stay on the filter for at least another 24 h, a period long enough for induction to be completed (Fig. 3.6B). No morphogenetic effects could be demonstrated in such experiments when the 'membrane preparation' was covered with competent metanephric mesenchyme.

An indirect way to explore the signal substances and their mode of action would be to use specific metabolic inhibitors. We have published extensive series of such experiments, but the conclusions as far as the mechanisms are concerned remain meagre. Actinomycin D, which in low concentrations inhibits transcriptional processes, completely blocked tubule formation at a concentration of 0.05 μg/ml. Separate exposure of either the mesenchyme or the spinal cord for 30 min prior to recombination had the same effect. Treatment of the recombinant explants after 24 h of cultivation no longer inhibited subsequent tubule formation

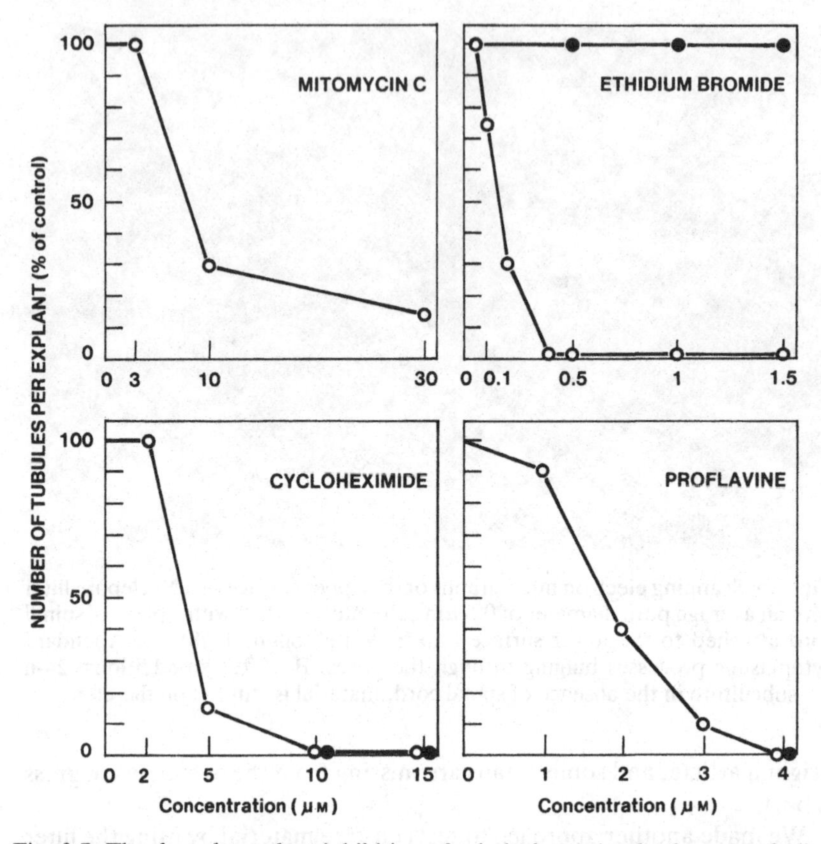

Fig. 3.7. The dose-dependent inhibition of tubule formation by four metabolic inhibitors when applied during the first 24 h of transfilter culture (after Nordling *et al.*, 1978). Open symbols, spinal cord removed; closed symbols, spinal cord not removed.

(Jainchill *et al.*, 1964). When inhibition of the incorporation of [^{3}H]uridine was measured at different stages of the experiment, a maximum inhibition of 50% was recorded at 24 h when a peak of RNA synthesis also had been observed (Vainio *et al.*, 1965). Similarly, the typical post-inductory shift in the lactate dehydrogenase pattern was prevented by actinomycin treatment during the early contact, but not after 30 h of transfilter culture (Koskimies, 1967*a*). The significance of these and many similar experiments is, however, lessened because we still do not know exactly the mode of action of the drug, which also affects the viability of the exposed cells.

Further experiments included exposure to various inhibitors of DNA, RNA and protein synthesis (mitomycin C, ethidium bromide, proflavine and cycloheximide) (Nordling *et al.*, 1978). A tubule-inhibiting concentration not severely affecting the viability of the cells was found with

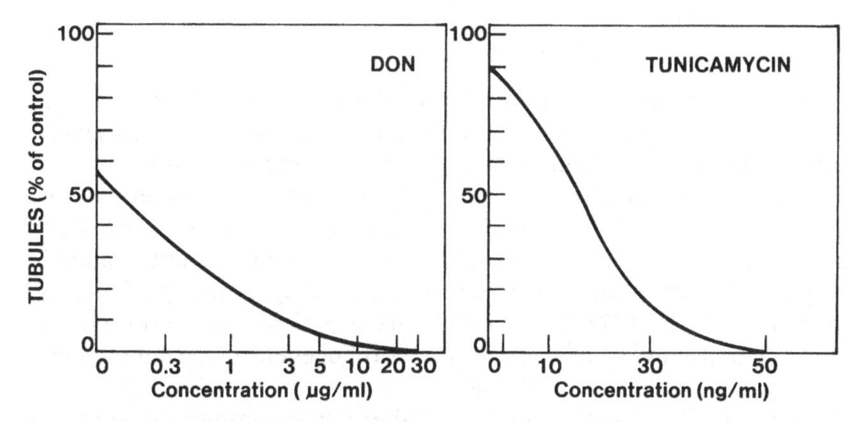

Fig. 3.8. Results of experiments where the metanephrogenic mesenchyme/spinal cord explants were treated with 6-diazo-5-oxo-norleucine (DON) or tunicamycin during the first 24 h of cultivation (after Ekblom *et al.*, 1979*a*, *b*).

all compounds, and it was associated with concentrations affecting synthesis of macromolecules (Fig. 3.7).

The glutamine analogue 6-diazo-5-oxo-norleucine (DON) interferes with the synthesis of glycosaminoglycans and glycoproteins and shows a dose-dependent inhibition of tubule formation in transfilter experiments (Ekblom *et al.*, 1979*a*; Fig. 3.8). The effect was obtained only during the critical induction period, the first 24 h of transfilter culture, after which the same concentrations were without detectable effect. Ekblom *et al.* conclude that carbohydrates, most likely at the inductor/target interphase, are associated with the actual induction process. Similar results were obtained after addition of tunicamycin, an inhibitor of protein glycosylation. The dose-dependent inhibition of induction during the critical 24 h (Fig. 3.8) correlated well with the inhibition of protein glycosylation measured by ^{3}H-labelled mannose (Ekblom *et al.*, 1979*b*). Only a slight reduction in the number of tubules was seen after the 24-h period. It might be concluded that impaired glycosylation during induction affects the process, but because both tissues were exposed it is not possible to distinguish between effects on the induction, on the target cells or on the transmission of signals.

To conclude, the extensive and lengthy experiments summarized above and in Figs. 3.7 and 3.8 suggest that impaired protein synthesis, protein glycosylation, and synthesis of carbohydrates interfere with subsequent tubule formation. This could be shown only during the induction period. When effective concentrations of the inhibitors were added after the initial stage, differentiation appeared to proceed normally. For the reasons given below, it is tempting though not fully justified to suggest that the critical molecules lie in the inductor/target interphase.

Transmission of inductive signals

As long as the chemical nature, localization and mode of action of the compounds apparently carrying inductive messages remain unknown, indirect information might be gained from studying the transmission of the message, i.e. the conditions where an induction can be achieved. The first in a long series of such experiments has already been mentioned: lens induction is prevented by the introduction of impermeable material between the optic cup (the inductor) and the epidermis (the target) (Spemann, 1901, 1912). Such interference experiments have since been performed in many model-systems for morphogenetic tissue interactions (for reviews, see Saxén, 1972, 1980).

Grobstein (1956*b*) postulated three alternative transmission mechanisms for inductive signals: (1) induction mediated by diffusible compounds, (2) induction via actual cell-to-cell contacts, and (3) interaction of compounds of the extracellular matrix. More recently, after experimental results in many model-systems, I have re-phrased the original postulate (Table 3.3). Among the alternatives presented in the table, free diffusion of signal substances and matrix-mediated interactions have been well established, the former during amphibian primary induction (Saxén, 1961; Toivonen, 1979) and during the induction of the lens (Karkinen-Jääskeläinen, 1978). A matrix-mediated interaction is the probable mechanism for the inductive events in tooth development (Thesleff *et al.*, 1977, 1978) and in epithelial–mesenchymal interactions in glandular organs (Bernfield *et al.*, 1972, 1984*a*, *b*). Short-range transmission of inductive signals will be discussed below, but the outlined alternative mechanisms 3 to 5 in Table 3.3 should still be considered as merely theoretical speculations without direct experimental proof. For exploring such short-range (contact-mediated) processes or to exclude their significance, two sets of results become relevant, those derived from transfilter experiments and those from direct ultrastructural and immunohistochemical examinations of the inductor/target interphase. The discussion will be restricted to the developing kidney.

In order to distinguish between the alternative transmission mechanisms, Grobstein (1956*a*) devised the transfilter technique described on p. 48 and illustrated in Fig. 2.9. Two types of Millipore filters, made of cellulose ester, were specifically manufactured for these experiments, the 20-μm thick HA filter, with a calculated pore size of 0.45 μm, and the somewhat thicker AA filter, with an average pore size of 0.8 μm. In the situation described, the AA filters regularly allowed passage of the inductive signal from the spinal cord to the mesenchyme. The HA filters allowed this passage less regularly, and the effect was weaker (a smaller number of tubules). Cytoplasmic processes were frequently found to penetrate filters with large pores, but rarely the HA filters with 0.45-μm

Table 3.3. *Alternative modes of transmission of inductive signals*

Long-range transmission (50 000 nm)
1. Free diffusion
2. Matrix interaction
Short-range transmission (5 nm)
3. Short-range diffusion
4. Interactions of surface-associated molecules
5. Transfer of molecules through intercellular channels

Data from Grobstein 1955*b*; Saxén *et al.*, 1976*a*; Weiss & Nir, 1979; Saxén, 1981.

pores that still allowed the passage of induction. The absence of cytoplasmic material in filters with a mean pore size of 0.1 μm that occasionally allowed weak induction was shown by electron microscopy, and the conclusion that 'inductive activity in this system is not dependent upon cytoplasmic contact and hence resident in materials which are at least potentially extra-cytoplasmic' seemed to be justified (Grobstein & Dalton, 1957). In fact, such material had already been seen by Grobstein (1956*a*), in and on the filters attached to the spinal cord. This trypsin-digestible material was located in areas where tubule formation was initiated, and it seemed to be a good candidate for the actual signal substance.

The transmission characteristics of tubule induction were further explored by Grobstein (1957) with filters of varying thickness – either a multi-layered thin filter assembly or a single thick filter. The increase of the inductor–mesenchyme distance led to a gradual decrease in the intensity of induction and to its complete blockage when the filter thickness reached approximately 80 μm. Response could also be estimated by measuring the thickness of the mesenchyme, as uninduced mesenchyme tends to spread as a thin layer on the filter (Fig. 3.9). In all, the experiments by Grobstein (1956*a*, 1957) suggest that kidney tubule induction is mediated by extracellular factors that travel over a distance of 60 to 80 μm without requiring actual contact between the interacting cells. Later, however, Grobstein (1961) emphasized that this is not a free diffusion which carries molecules from the inductor to the target cell over a 'free space', but rather extracellular components with restricted mobility. Induction might be regarded as taking place when two cells are in 'juxtaposition which brings into interaction the microenvironments of cells and tissues'.

The above conclusion was re-evaluated after observations on the kinetics of the inductive process. Both Grobstein and we had come to the conclusion (unpublished) that induction in the transfilter condition is rather slow, and that it is not completed until the second day (see later).

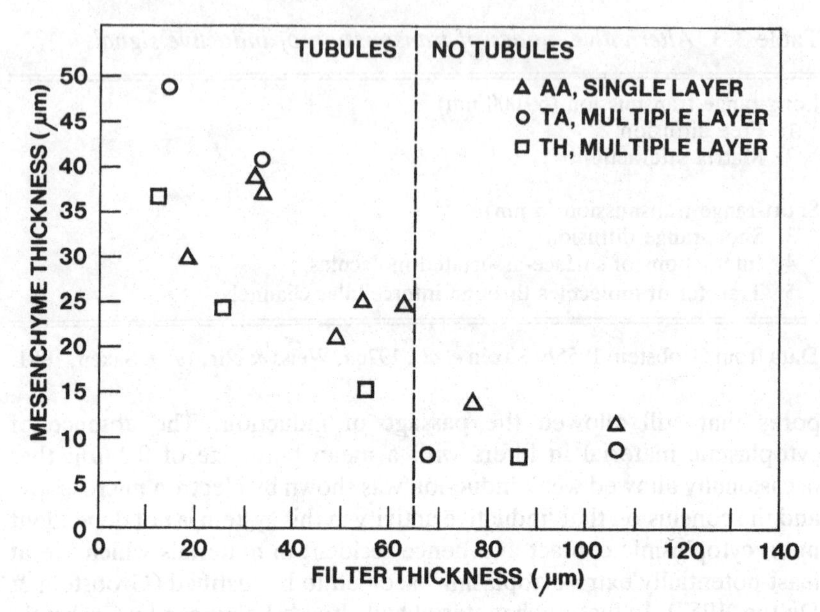

Fig. 3.9. The relation between the thickness of the mesenchyme and formation of tubules in cultures where the metanephric mesenchyme was separated from the spinal cord by a filter layer of varying thickness (after Grobstein, 1957). AA, thick filter, pore size 0.8 μm; TA, thin filter, pore size 0.8 μm; TH, thin filter, pore size 0.45 μm.

This period apparently consists of several time-consuming events (Grobstein, 1967): (1) the production of the signal substance, (2) the actual transfer, and (3) the response of the mesenchyme. To these events can be added (4) adaptation time to recover from the initial, experimental manipulations (Vainio *et al.*, 1965). To distinguish between the actual transfer time and the other components of the 'minimum induction time', the total contact time required for the induction to be completed was measured in single and double filter assemblies. The postulate was that the adaptation time, production time, and response time would be the same in both series, whereas a possible prolongation of the total time in the double-filter experiments should reflect the time needed for the 'inductor' to pass through the second filter. The results (Fig. 3.10) show that in the single filter series, the induction was completed during the first 24 h, whereas the second filter prolonged the time by some 12 h. This slow transfer of the message did not obey the rules of passive diffusion, and actual measurements of the transfilter passage of various macromolecules revealed that the traverse was much faster. Hence mechanisms other than free diffusion should be considered for the passage of the tubule-inducing stimulus (Nordling *et al.*, 1971).

Subsequently, three types of thin Millipore filters with average pore

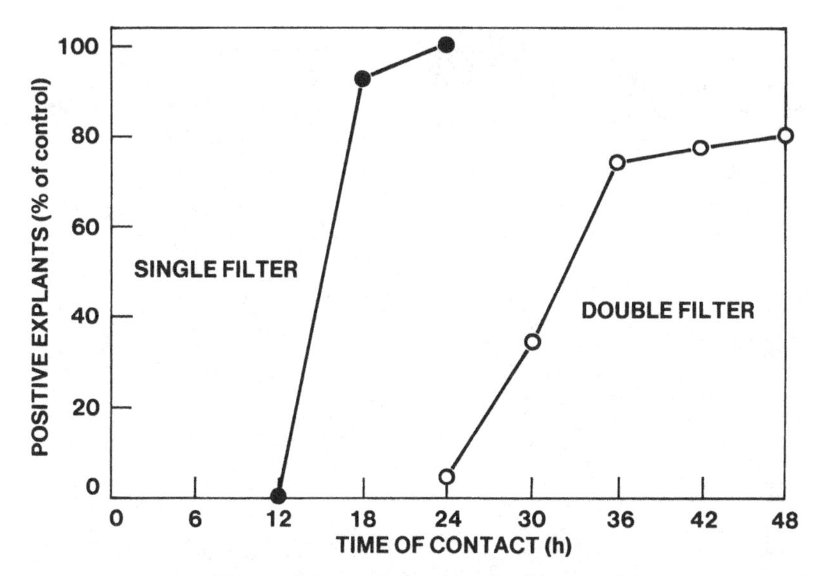

Fig. 3.10. Scheme of the 'minimum induction time' in transfilter cultures where either a single or a double 28-μm thick Millipore filter was interposed between the metanephric mesenchyme and the spinal cord (after Nordling *et al.*, 1971). The technique is shown in Fig. 3.16.

sizes of 0.1, 0.2, and 0.8 μm were used. Induction occurred with 0.2- and 0.8-μm filters, but with the 0.1-μm filter it was obtained only occasionally. Light microscopy of thin Epon-embedded sections revealed abundant cytoplasmic material in the two filters that had allowed induction. In the 0.1-μm filters there was only shallow ingrowth. After improved fixation, electron microscopy showed cytoplasmic material in them also (Fig. 3.11). It was concluded that a filter with the smallest pore-size did not necessarily eliminate close apposition of cell processes, and mechanisms based on 'cell contacts' should still be considered.

The introduction of the Nuclepore filters in the early 1970s provided a good new tool for our studies. This polycarbonate filter has pores of known and rather uniform sizes, and the channels run relatively straight through the membrane (Fig. 3.12). The filters were used in comparisons of the penetration of cytoplasmic material through the filter with the passage of induction (Wartiovaara *et al.*, 1972, 1974; Lehtonen, 1976; Saxén *et al.*, 1976*b*). Results of these studies are summarized in Table 3.4 and illustrated in Figs. 3.13 and 3.14. In conclusion, the penetration of cytoplasmic processes through the filters is associated with the passage of the inductive signals. We consider this to be strong though circumstantial evidence for transmission, based on a close apposition of the interacting cells. The results do not distinguish between the possibilities 3 to 5 in Table 3.3. Very little, if any, material was detected between the

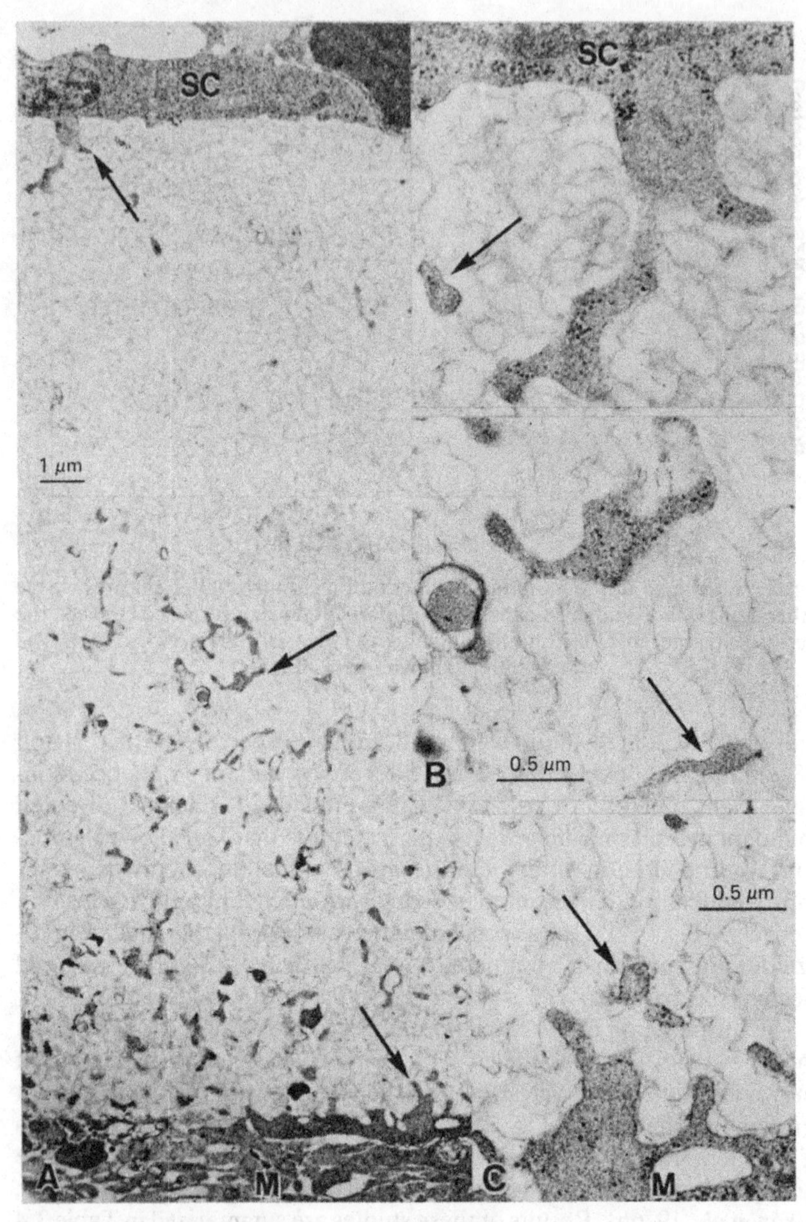

Fig. 3.11. Electron micrographs of a Millipore filter with an average pore size of 0.1 μm interposed between the metanephric mesenchyme and the spinal cord. Cytoplasmic material and membrane-coated processes (arrows) are seen in the filter pores (Lehtonen *et al.*, 1975; courtesy of Dr E. Lehtonen). SC, spinal cord; M, mesenchyme. A. General view of the filter with the two tissues attached. B. View from near the spinal cord. C. The same filter examined at the filter/mesenchyme contact surface showing cytoplasmic extrusions into the filter.

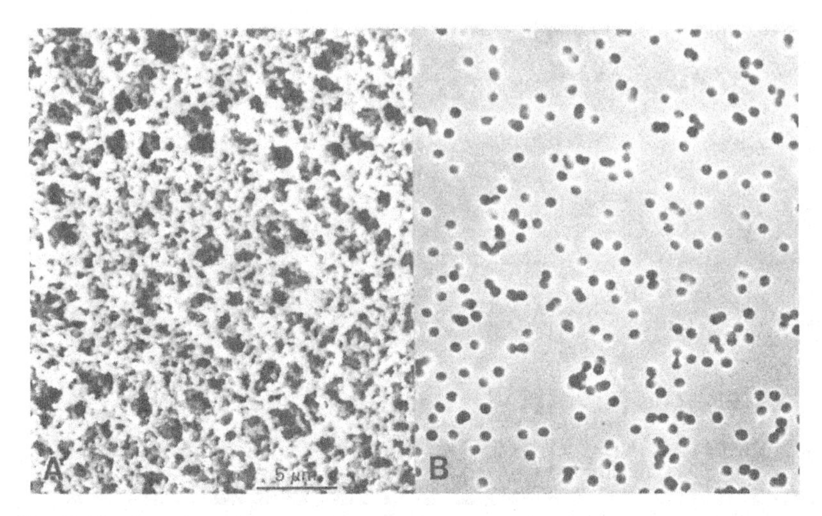

Fig. 3.12. Scanning electron micrographs comparing the surface structure of a Millipore filter (average pore size of 0.8 μm) (A) with a Nuclepore filter (pore diameter of 0.5 μm) (B) (Lehtonen, 1975; courtesy of Dr E. Lehtonen).

Table 3.4. *Transfer of the tubule-inducing stimulus through Nuclepore filters with varying pore sizes as compared with the penetration of cytoplasmic material evaluated in electron microscopy*

Filter: nominal pore size (μm)	Inductor: Salivary mesenchyme		Inductor: Spinal cord	
	Tubules	Penetration	Tubules	Penetration
0.05	n.d.	n.d.	0/25	–
0.1	n.d.	n.d.	3/30	+
0.2	0/13	–	21/31	++
0.6	0/10	±	25/25	+++
3.0	8/14	+	15/15	+++

Two heterologous inductors, spinal cord and salivary gland mesenchyme were tested (Saxén *et al.*, 1976*b*; Saxén, 1980).
n.d., not determined.

intra-filter processes of the two tissues (Fig. 3.14C,D), and no specialized membrane junctions (channels) were found.

After developing a rough quantitative method to evaluate the strength of the induction, we examined the significance of filter porosity, i.e. the extent of contact (Saxén & Lehtonen, 1978). This could be explored after the manufacturer gave us two types of filters with the same pore size but

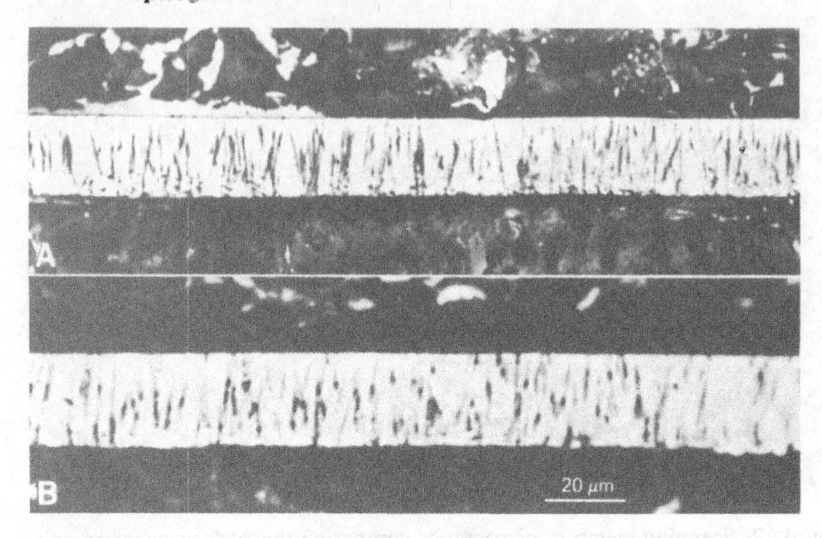

Fig. 3.13. Micrographs of Nuclepore filters with average pore sizes of 0.2 μm (A) and 0.6 μm (B), showing abundant cytoplasmic material in the pores when grafted for 24 h in contact with metanephrogenic mesenchyme and spinal cord (below the filter) (courtesy of Dr E. Lehtonen).

Table 3.5. *Induction of kidney tubules in the metanephric mesenchyme through Nuclepore filters of 'low' and 'high' porosity (average pore size 0.5 μm)*

Porosity (pores/cm²)	Positive cases	(%)	Mean number of tubules per positive explant	per explant
2×10^6	(n = 27)	52	7.8	3.7
2.3×10^7	(n = 20)	100	20	20

From Saxén & Lehtonen, 1978. *n*, number of explants.

with a definite difference in the actual porosity (pores/area). The results (Table 3.5) show a clear 'dose-dependence' of the induction.

The epithelial–mesenchymal interphase

If the above conclusions on the role of cell contacts mediating induction are accepted, such a close apposition of the interacting epithelial and mesenchymal cells should be detected in the normal situation *in vivo*. An electron-microscopic analysis was therefore performed by Lehtonen (1975). A basement membrane was regularly detected around the inductively inactive stalk of the branching ureter, but it was discontinuous at

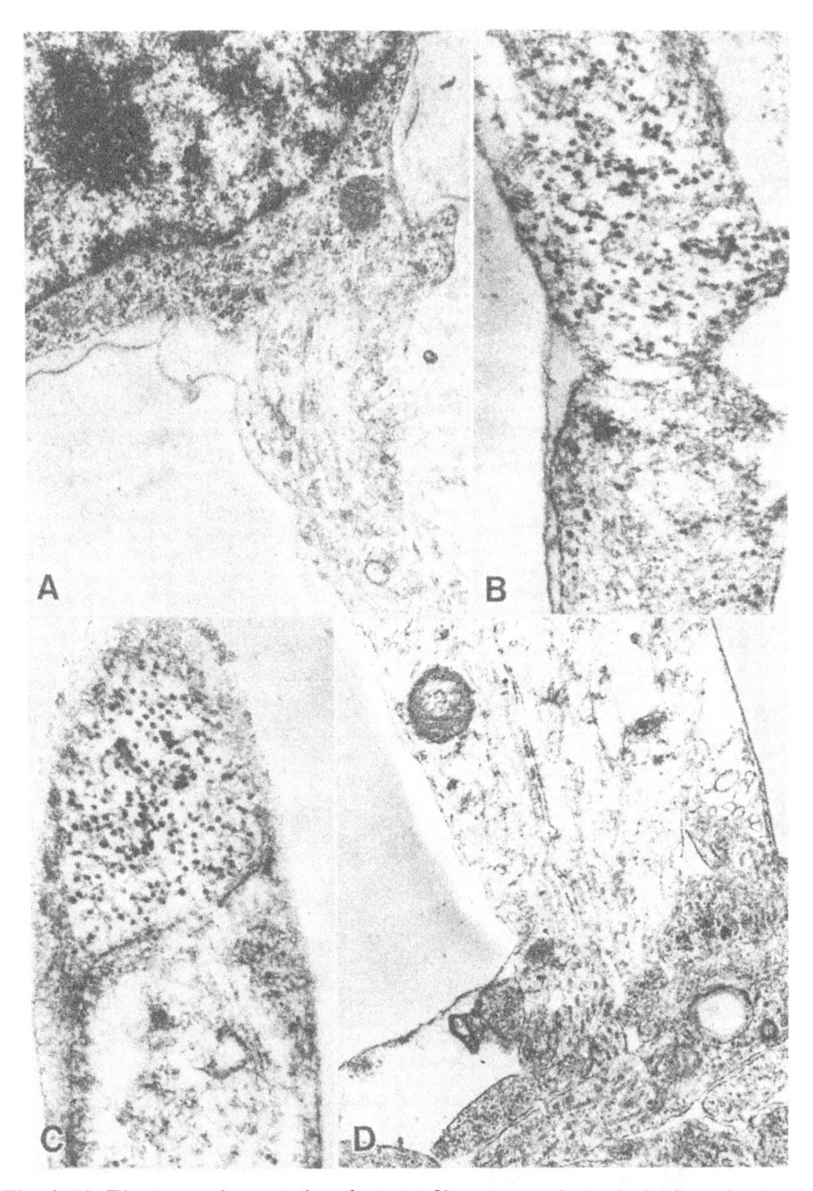

Fig. 3.14. Electron micrographs of a transfilter mesenchyme/spinal cord culture with an interposed Nuclepore filter, showing cytoplasmic penetration from the mesenchyme (A) and from the spinal cord (D). Close apposition of the membranes is seen within the filter pores (C and D) (Wartiovaara *et al.*, 1974; Lehtonen, 1976; courtesy of Dr J. Wartiovaara and Dr E. Lehtonen).

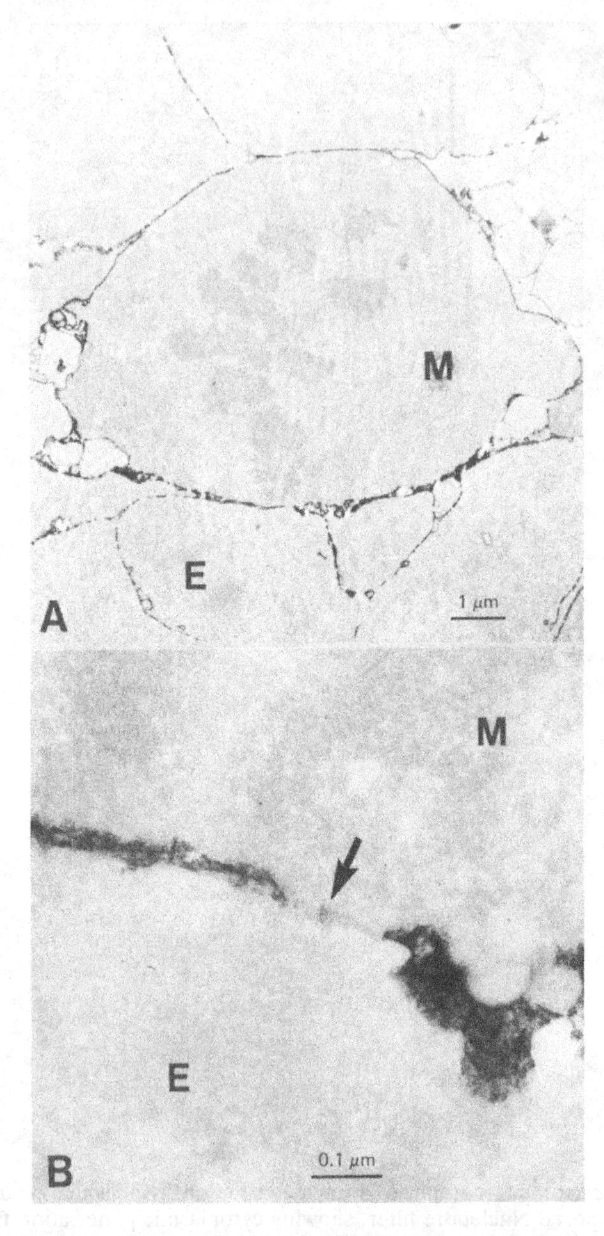

Fig. 3.15. Electron micrographs of the epithelial/mesenchymal interface in an embryonic kidney. Ruthenium red–uranyl acetate staining (Lehtonen, 1975; courtesy of Dr E. Lehtonen). A. Ruthenium-red-positive material is seen around the epithelial (E) and mesenchymal (M) cells and in their interface. B. Apparent focal fusions are seen at points between the epithelial and mesenchymal cells (arrow), devoid of the ruthenium-red-positive material.

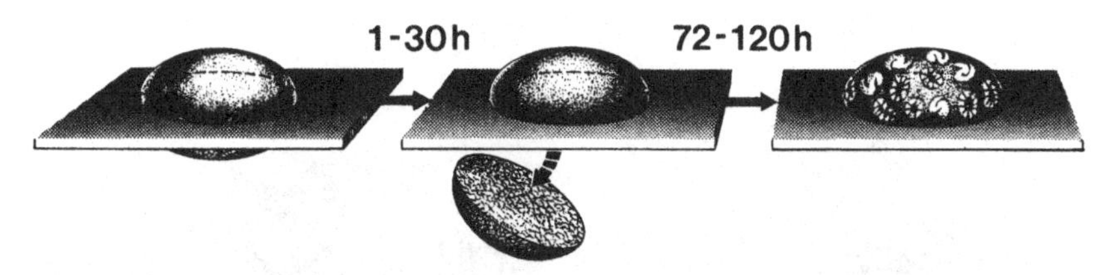

Fig. 3.16. Scheme of the transfilter experiment to study the kinetics of induction. After various intervals the inductor is scraped off and the mesenchyme subcultivated on the filter.

the active tips. The epithelial and mesenchymal cells came to a close apposition, and the intercellular gap was less than 10 nm. Very little ruthenium-red-positive extracellular material was found here (Fig. 3.15), and Lehtonen concludes that the findings are in agreement with the conclusions of the transfilter studies and 'obviate the need to postulate long-range transmission of inductive signals'. In our later studies this lack of the basement membrane at the tips of the ureter could be visualized by immunohistology with antibodies against the known basement membrane components laminin and type IV collagen.

Kinetics of the tubule induction

As shown in Fig. 3.10 and suggested earlier by Grobstein (1967), the transfilter induction is a time-consuming event that requires some 24 h of contact to be completed in this experimental situation. The new Nuclepore filters allowed us to examine the matter more closely, and the basic method is schematized in Fig. 3.16: after the chosen time, the inductor (spinal cord) was carefully scraped from the lower surface of the filter, and the mesenchyme was subcultured for various periods. The 'normal' morphogenesis of the mesenchyme in these conditions with the inductor left in place is described on pp. 87–91. Accordingly, at 24 h, no visible changes have yet occurred and only after a 12- to 24-h subculture can clear condensates be found. Renal vesicles are seen somewhat later. Consequently, in the following experiments, the total culture period was 72 h, and the tubules were counted from material fixed at that time (Saxén & Lehtonen, 1978; Fig. 3.17). The results obtained with two types of Nuclepore filters with different pore sizes are illustrated in Fig. 3.18. They show that the minimum induction time is a function of the pore size. When relatively large pores (0.6 μm) are used, the first cells to become irreversibly converted into epithelial tubules are determined after 12 h of transfilter contact, and the induction is completed around 24 h. The same time-course was later obtained for filters of 1.0 μm, whereas the minimum time was prolonged when filters with 0.2 μm pores were used.

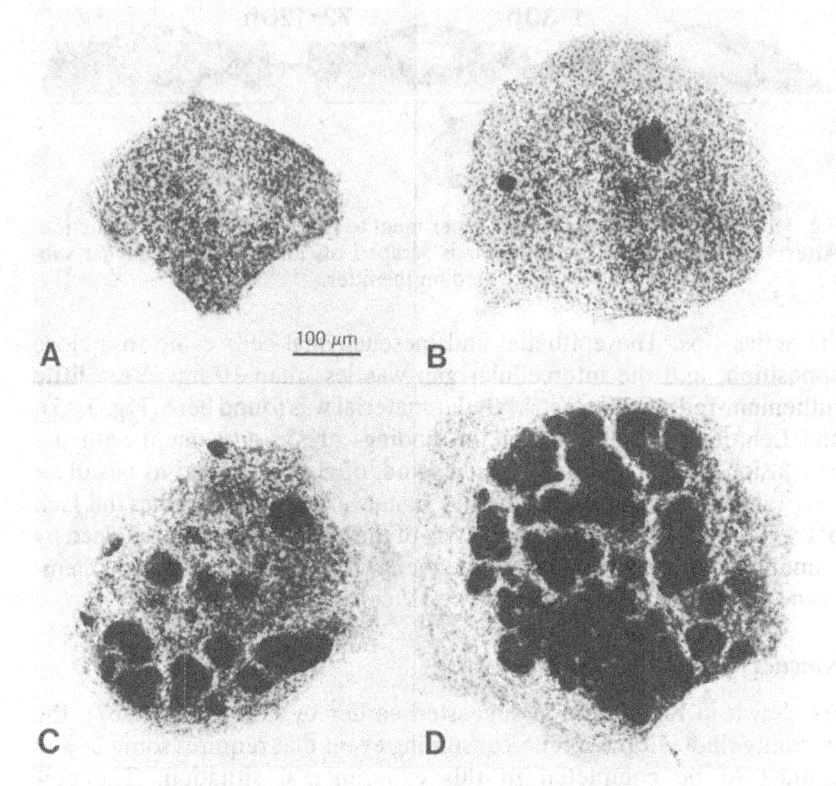

Fig. 3.17. Micrographs of whole mount preparations made from mesenchymes exposed to the inductor for various periods (Fig. 3.16) and subcultivated for 72 h (Saxén & Lehtonen, 1978). A, uninduced mesenchyme; B to D, mesenchymes induced for 12, 18 and 24 h, respectively.

Our results diverge somewhat from those of Gossens & Unsworth (1972). These authors used Millipore filters with the nominal pore size of 0.45 μm and estimated the minimum induction time to be slightly longer than in our experiments, i.e. 30 h. They recorded regression of the early tubules when the inductor was removed and therefore suggested a two-step process with an initial induction followed by a maintenance state mediated by exogenous factors. Probably these 'second-step requirements' of Gossens & Unsworth were merely permissive, nutritional conditions, as heterologous mesenchymes and embryo extract both showed this action.

The minimum induction time tested with Nuclepore filters is of the same order of magnitude as with the spongy Millipore filters, i.e. approximately 12 h for the first cells and 24 to 26 h for the induction to be completed. The 12-h prolongation when a double filter is used should reflect the time required for the message to pass through the second filter.

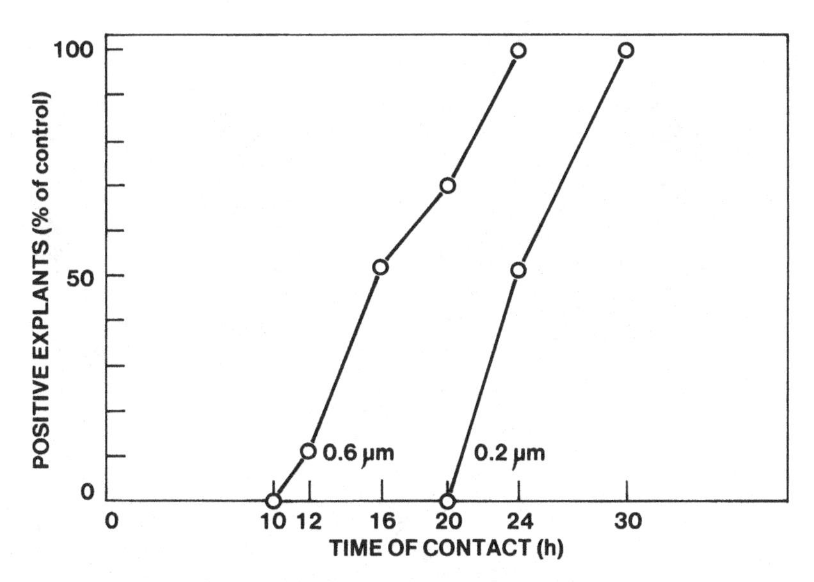

Fig. 3.18. The results of an experiment according to the scheme in Fig. 3.16, where the time-dependent increase of induced tubules and the association to the size of the filter pores are shown (after Saxén & Lehtonen, 1978).

It is not clear what occurs in the system during this rather long period. Since ingrowth of cytoplasmic processes is known to be rapid (1 to 2 h, Saxén & Lehtonen, 1978), other factors should be considered. A slow transport of material along the long, thin filaments might explain the diameter-dependent induction time (Fig. 3.18). Resynthesis of membrane-associated compounds at the contact surfaces might be another explanation.

Spread of induction in the target tissue

If one believes, as we do, that induction of kidney tubules is implemented by short-range, contact-dependent transmission of the message, another problem has to be faced: the induction is not limited to the cells next to the inductor, but the 'induction field' has a depth of several cell layers. This can be seen *in vivo* but perhaps more clearly *in vitro*. When the spinal cord and the mesenchyme are combined, the target cells spread deep into the mesenchyme, far from the inductor–mesenchyme border (Fig. 3.19). Similarly, in the transfilter experiments, cells are seen all the way up to the upper layers of the mesenchyme.

In order to test the actual width of the 'induction field', i.e. the distance of the most peripheral epithelial cells from the inductor/target interphase, the transfilter assemblies illustrated in Fig. 3.20 were used (Saxén & Karkinen-Jääskeläinen, 1975). The amount of target mesenchyme was

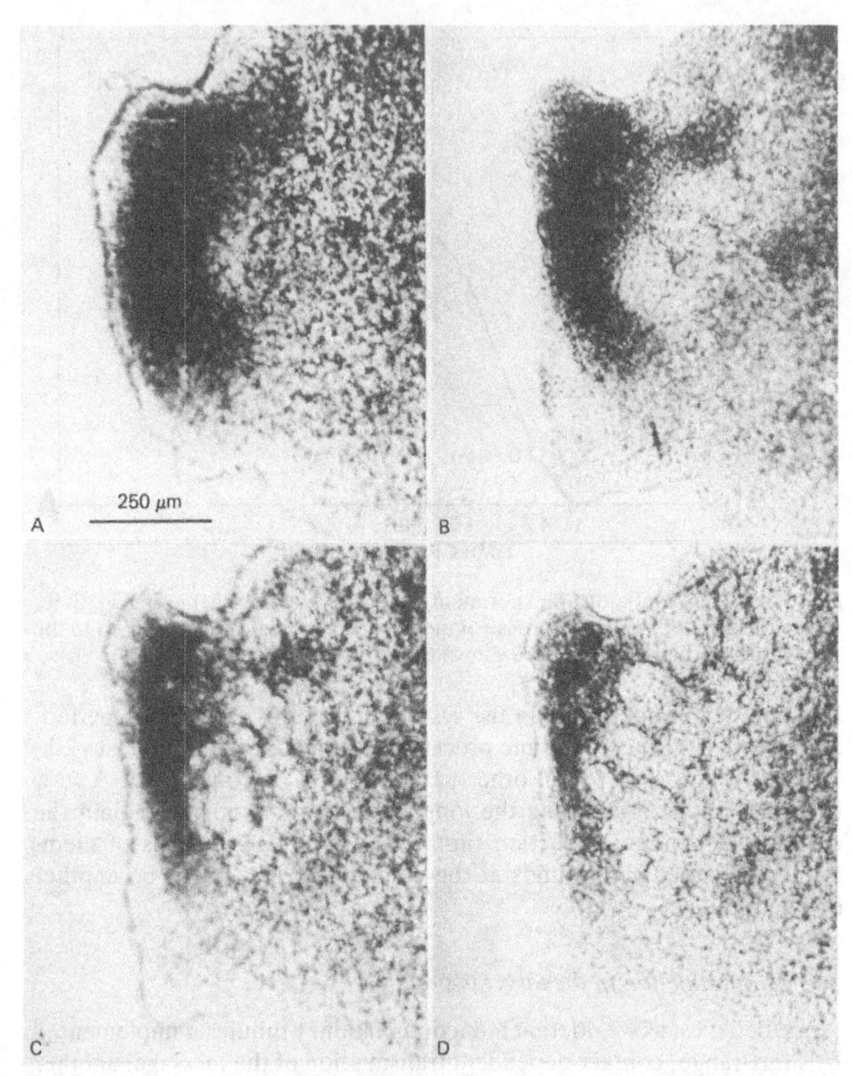

Fig. 3.19. A to D. Micrographs of a living mesenchyme/spinal cord recombinant, showing the gradual spread of response in the target tissue. Prints were made at 24-h intervals (Saxén & Wartiovaara, 1966).

increased in thickness by a 'tower' of mesenchymes built between supporting filters, or the lateral spread was measured as shown in the figure. After a period in culture of three to four days, the explants were sectioned serially and the distance between the tubules and the filter surface was measured. Tubules were seen regularly at a distance of 100 μm and occasionally as far as 150 μm from the filter surface.

Since long-range diffusion of an inductive signal was excluded in the

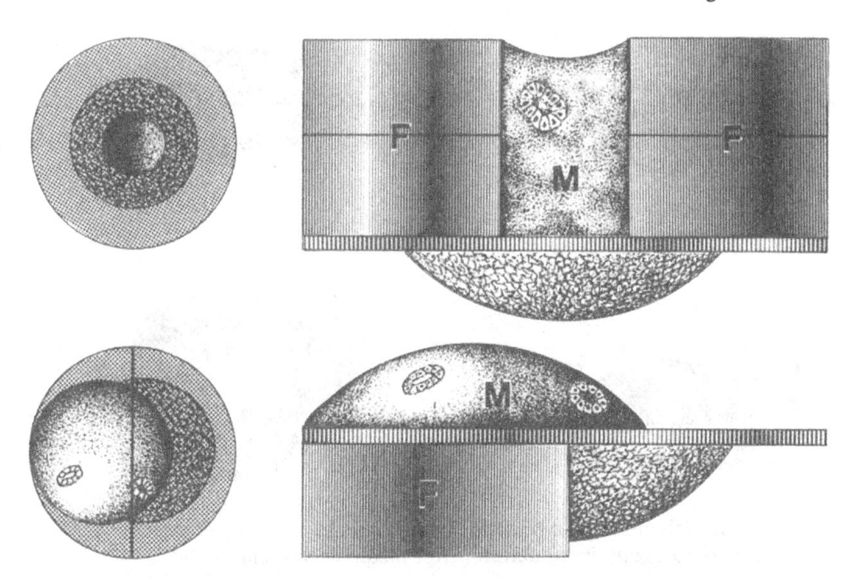

Fig. 3.20. Scheme of two modifications of the transfilter technique devised to estimate the transmission distance of the tubule-inducing message in the metanephric mesenchyme (M). A thin, 25-μm Millipore filter separates the interactants while thick, 150-μm filters (F) are used for supporting the structures (after Saxén & Karkinen-Jääskeläinen, 1975).

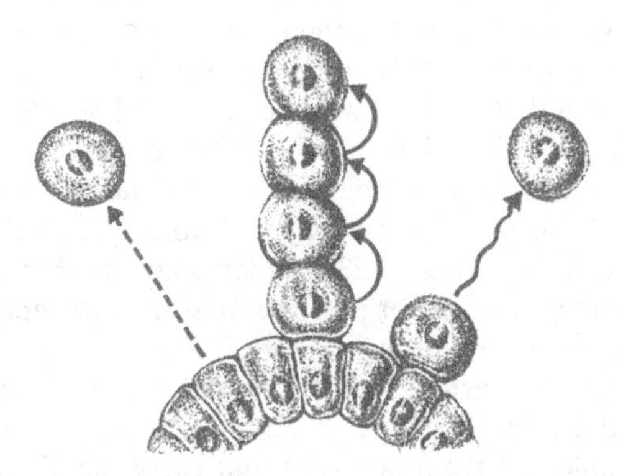

Fig. 3.21. Scheme of the alternative mechanisms for the spread of the inductive 'wave' in the target tissue.

transfilter experiments, two spreading mechanisms remained to be explored (Fig. 3.21). A homoiogenetic induction, i.e. a passage of the inductive effect from one induced cell to the next in a chain-like reaction, has been suggested to be involved in some other induction systems (Cooper, 1965; Deuchar, 1970). Another mode of spreading would be the migration of the induced cells from the inductor/target border into

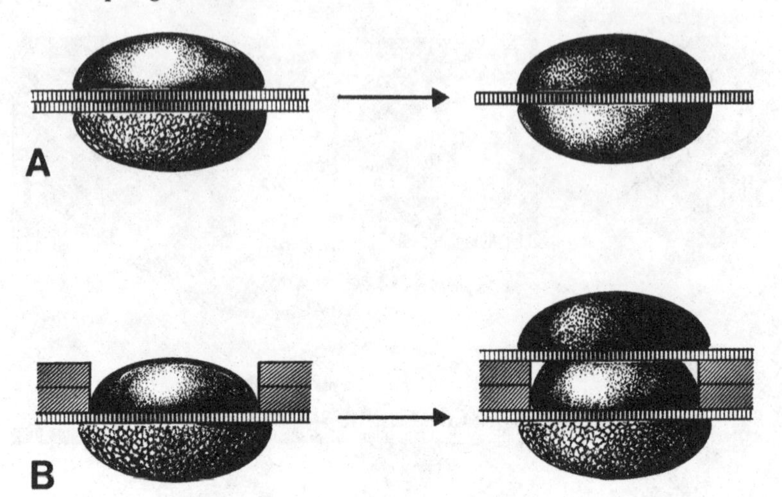

Fig. 3.22A, B. Scheme of the transfilter experiments made for examination of a homoiogenetic induction (after Saxén & Saksela, 1971).

deeper layers of the mesenchyme. The two alternatives were tested in a series of transfilter and recombination experiments.

Figure 3.22 illustrates the transfilter assemblies used for testing the hypothesis of a homoiogenetic induction: a metanephric mesenchyme was induced through a double filter, and then the contact was broken between the two filters (to avoid contamination of cells as described on p. 63). Thereafter a second set of competent mesenchymes was added to the clear filter surface and the explant was subcultured (Fig. 3.22A). Another design (Fig. 3.22B) was based on previous experience that the effect from the spinal cord cannot travel all the way through the interspace between the inductor and the upper mesenchyme. Both types of experiments yielded negative results (Fig. 3.23), which suggests that a transfilter-induced mesenchyme would not pass the message onto another mesenchyme in the transfilter situation.

Since the above experiments were based on the rather artificial transfilter situation, another type of design was also used (Fig. 3.24). Mesenchymes collected from embryos that carry the T_6 chromosomal marker were initially induced for 24 h in a transfilter setup. When induction was completed, the inductor was removed, the mesenchymes on the filter surface were cut into pieces, and fragments of freshly isolated mesenchymes of normal karyotype were added (at this stage no condensates had formed). After a subcultivation of 72 h, the mesenchyme pieces were collected and the tubules enzymically separated and cleaned. Monolayer outgrowths were then prepared from these tubules, and the karyotype of the cells was determined after colcemid treatment. If a homoiogenetic induction had occurred and the message had passed on

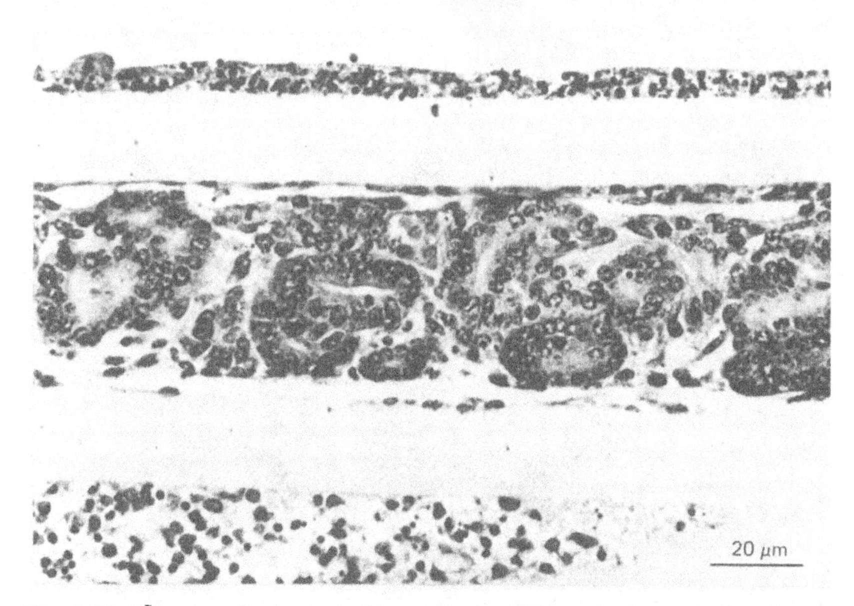

Fig. 3.23. Micrograph of a transfilter culture of the type shown in Fig. 3.22B. Spinal cord (below) has induced tubules in the mesenchyme on the opposite side of the filter, but induction has not been carried onto the second mesenchyme (on top) (Saxén & Saksela, 1971).

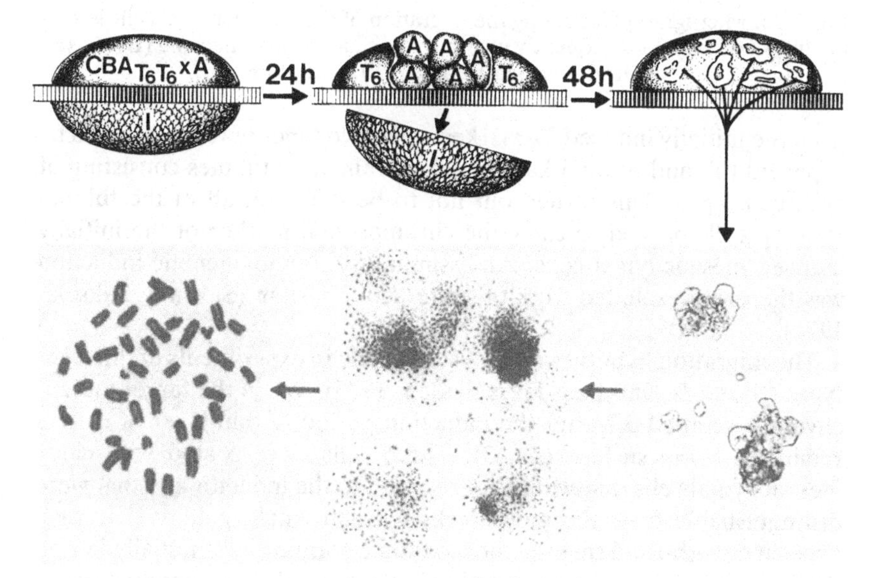

Fig. 3.24. Scheme of the experimental design to study homoiogenetic induction by the use of a chromosomal marker (after Saxén & Saksela, 1971).

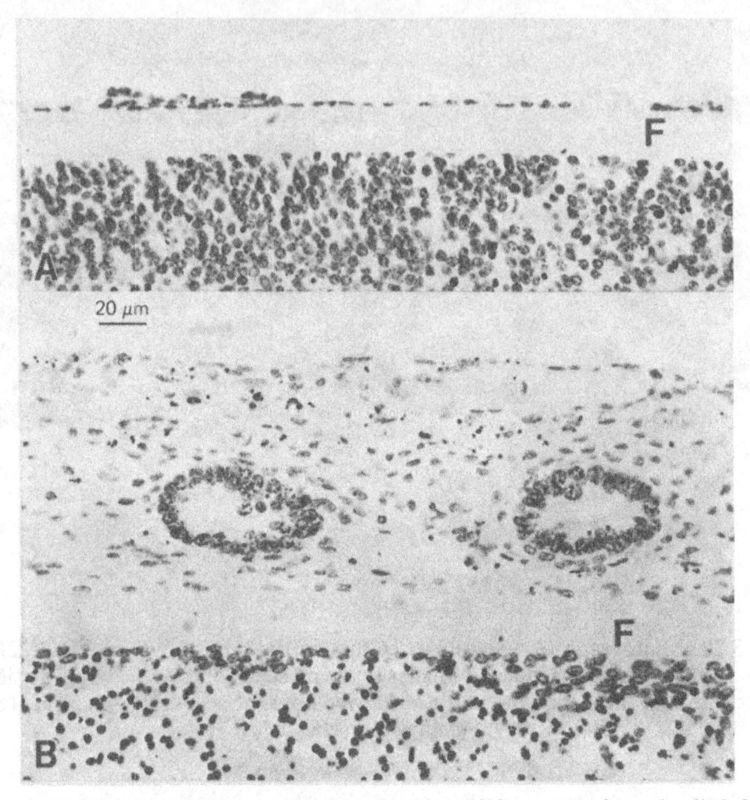

Fig. 3.25. Micrograph illustrating the migration of the mesenchymal cells left on the filter (A) into a subsequently overlaid chick gastric mesenchyme (B) (Saxén & Karkinen-Jääskeläinen, 1975). F, filter.

from the initially induced T_6-marked cells, two types of tubules would be expected (T_6 and normal karyotype) or chimaeric tubules consisting of both cell types. This turned out not to be the case; all of the tubule-forming cells proved to carry the chromosomal marker of the initially induced mesenchymal cells. An assimilatory, homoiogenetic induction was therefore excluded from this interactive system (Saxén & Saksela, 1971).

The migration hypothesis was tested finally in experiments of different types (Saxén & Karkinen-Jääskeläinen, 1975). When the target mesenchyme is scraped off after the induction period, a thin layer of cells is retained on the surface (p. 63). These cells were coated with chick mesenchymal cells that would not respond to the inductor and that were distinguishable from the mouse cells by their small, pale nuclei. Such experiments showed that the mouse tubule-forming cells initially left on the filter surface could be found deep in the heterologous mesenchyme at a distance of between 30 to 100 μm, occasionally up to 150 μm from the

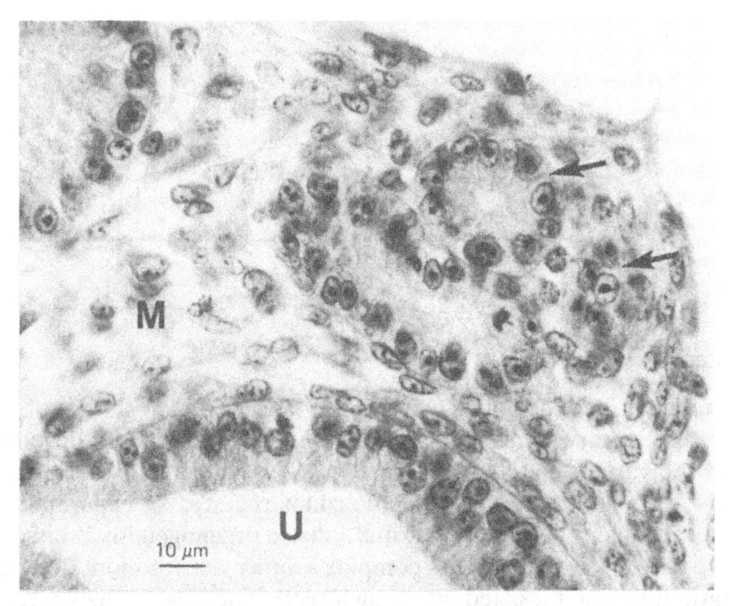

Fig. 3.26. Micrograph of a Feulgen-stained section from a chimaeric recombinant of quail ureteric bud (U) (with attached mesenchymal cells) and chick mesenchyme (M). The tubule-forming quail cells (nuclear marker) have migrated deep into the chick mesenchyme (arrows) (Saxén & Karkinen-Jääskeläinen, 1975).

filter surface (Fig. 3.25). The transfilter situation might again mimic normal development poorly, and hence an experiment was made with a chick/quail chimaeric-tissue combination. The quail cells are easily distinguishable from the chick cells by the 'quail nuclear marker' (Le Douarin & Barq, 1969). Quail ureters were dissected in the usual way, but not enzymically cleaned, and a thin layer of mesenchymal cells was left on the surface. Such ureters were combined with chick mesenchymal tissues. Quail-type tubules developed in these chimaeric explants, and they were frequently found at a distance from the original location of the cells, the ureteric surface (Fig. 3.26). The observation gave further, rather convincing support to the migration hypothesis.

In conclusion, induction of kidney tubules initially mediated by cell contacts is spread in the target mesenchyme by a peripheral migration of the induced cells. This would inevitably bring new, uninduced cells into contact with the inductor, and these would, in turn, become induced. Such a random movement can be visualized in time-lapse motion pictures of whole-kidney rudiments (Saxén *et al.*, 1965*a*). This mechanism could explain the rather slow induction process in transfilter contact: at 12 h the first cells in contact would be induced and transferred to the deeper layers, followed gradually by other similarly behaving cells, until the majority of them have become triggered at 24 h.

4

Experimental tubulogenesis

Introduction

Chapter 1 described the major features of the early organogenesis of the vertebrate excretory system and the complex, yet synchronous, development of the various cell lineages involved. A detailed analysis of these various events and their causal relationships requires an experimental approach in which the prolonged, multiphasic organogenesis is dissociated into single – or at least less complex – processes for more detailed exploration. Chapter 2 listed many such techniques and model-systems including cultivation of nephric tissues and their subfragments *in vitro*, various culture methods for kidney-derived cell lineages, and grafting of nephric material onto heterotopic sites such as avian chorioallantoic membrane. Probably the most advantageous technique is still the one devised by Grobstein (1956*a*, 1957), who utilized thin membrane filters. As described in detail on pp. 48–50, the transfilter technique involves an experimental triggering of the responding nephric mesenchyme towards epithelial direction, ultimately leading to the formation of nephric tubules. By this strictly controllable technique the differentiation of the determined target cells can be followed at molecular, cellular, and tissue levels. The events can be temporally correlated, which may allow us to draw conclusions as to their causal relationships. Consequently, Chapter 4 will be devoted to observations of cultured and experimentally induced metanephric mesenchyme. In addition to the advantage of exact timing, this technique and the corresponding model-system offer an adequate control tissue: a separated metanephric mesenchyme identically treated and cultivated but not exposed to an inductor is a proper control for most experimental situations. Needless to say, like all systems *in vitro*, the transfilter model-system is susceptible to many artifacts and does not necessarily mimic development *in vivo*. Some obvious limitations and fallacies will be pointed out in due course.

Early morphogenesis

The early morphogenetic changes of the metanephric mesenchyme experimentally induced and cultivated *in vitro* have been examined and described by Wartiovaara (1966*a*, *b*). These light- and electron-microscopic observations have been further reviewed by Saxén & Wartiovaara (1966) and by Saxén *et al.* (1968).

The starting material, the metanephric mesenchyme of an 11-day mouse embryo, shows no morphological signs of differentiation towards epithelial direction. The mesenchyme consists of irregularly shaped, non-polarized cells with randomly distributed pseudopodic protrusions. The distance between the contacting cell membranes is 100 to 150 nm, but wider gaps between the cells are common. By all morphological criteria, there seems to be only one cell type in the uninduced, undetermined mesenchymal blastema of the metanephros. The cells are uniformly distributed, and no regional differences in their density or assembly can be detected.

After dissection and cultivation *in vitro* for some 10 h the appearance of the nephric mesenchyme is retained whether cultivated transfilter with an inductor or kept isolated on a membrane filter. The first detectable difference is the flattening of the uninduced mesenchyme, which, after a total culture period of approximately 30 h, leads to a single-layered sheet of cells. In the induced mesenchyme, some regional differences are visible at around 10 h, when cells nearest the filter surface become packed into a 'basal layer' and show polarization, with their long axes oriented perpendicularly to the filter surface. Cells above this layer are still unpolarized and show no signs of differentiation or specific spatial assembly.

During the second day of transfilter contact with the inductor, the first pretubular aggregates form as the 'basal layer' splits into separate condensates (Fig. 4.1). These aggregates consist of non-polarized, irregularly shaped cells with decreased intercellular gaps. They soon become clearly delineated from the surrounding, uncondensed mesenchyme, and the intercellular gaps diminish further (Fig. 4.2). The cells within the aggregate assume a funnel shape, and a lumen is formed in the central portion of the cluster, first as a multifolded slit and soon after as a wider, irregular cavity. Polarization of the cells is also seen as a basal transposition of the nuclei (Figs. 4.3 and 4.4). A ring of junctional complexes joins the apical portions of the cell membranes, and abundant small, apparently Golgi-derived vesicles fuse to the juxta-apical and lateral cell membranes (Fig. 4.3). This stage is reached towards the end of the second day of transfilter cultivation, though development is not fully synchronous and different stages of differentiation are detected in the renal vesicles within the same mesenchyme.

Concomitant with the polarization of the now wedge-shaped epithelial cells and the appearance of the first signs of lumen formation, the peripheral protrusions disappear and basement-membrane-like extracellular material is deposited along the basal surface of the cells (Wartiovaara, 1966*a*; Bernstein *et al.*, 1981; Ekblom *et al.*, 1981*b*)

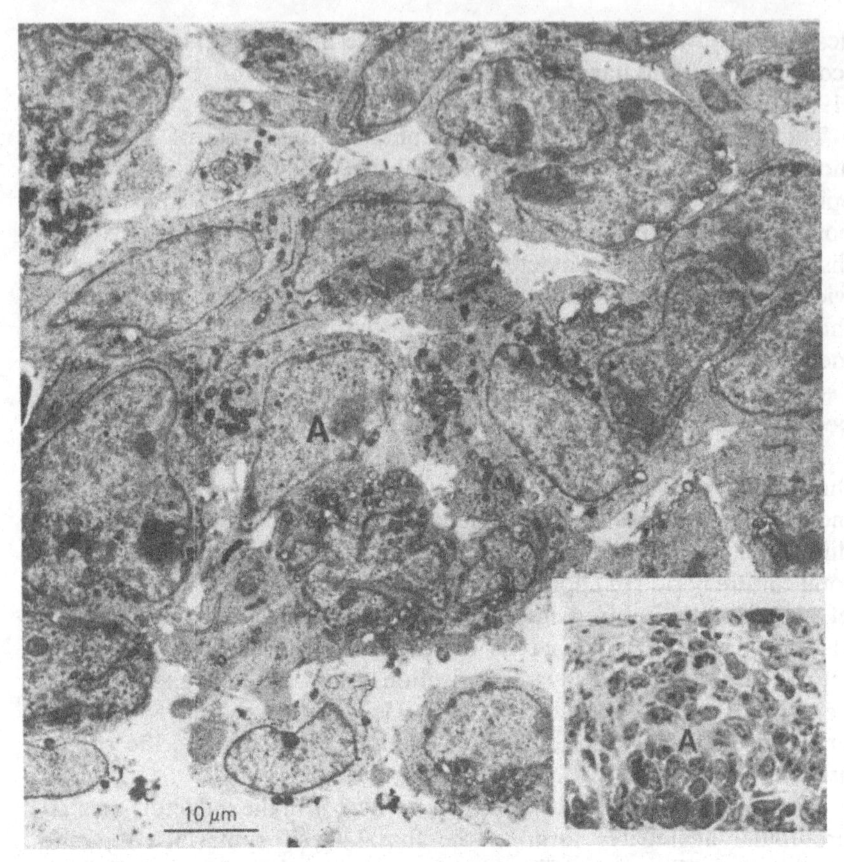

Fig. 4.1. Electron micrograph and light micrograph (inset) illustrating the initial stage of aggregate (A) formation in a transfilter-induced nephric mesenchyme. Within the aggregate, intercellular spaces become diminished but as yet no polarization of cells can be demonstrated (Saxén & Wartiovaara, 1966).

(Fig. 4.5). This material has been characterized further by immunohistological and biochemical means, and the result will be summarized below.

While the above-mentioned early morphogenetic changes in the mesenchymal cells bear a close resemblance to those taking place *in vivo*, the actual shaping of the tubules remains incomplete. Coiling of the tubules might be seen in prolonged cultures, but typical, regular S-shaped bodies are exceptions. Apparently some additional morphogenetic guiding forces are lacking *in vitro*. The inductor under the filter is stable as compared with the constantly moving and invading normal inductor, the ureter, which was postulated to create a gradient of adhesion in the remaining vesicle (p. 33). Another factor would be an interaction between the primitive renal vesicle and the stromal mesenchyme and/or certain soluble fractions, as suggested by Gossens &

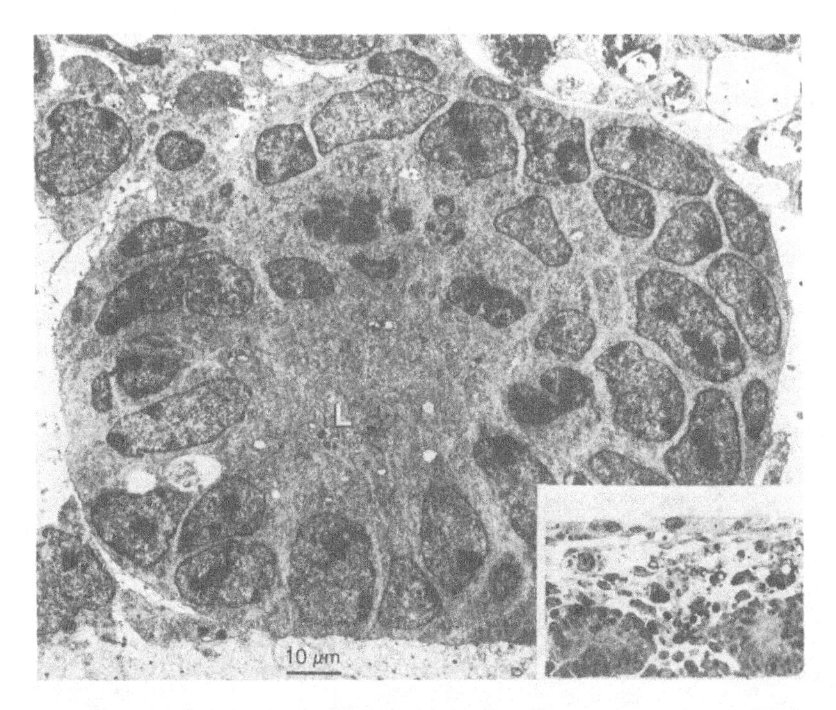

Fig. 4.2. Electron micrograph and light micrograph (inset) of an advanced stage of pretubular aggregate with distinct borders and scanty intercellular space. The beginning of orientation of the epithelial cells within the aggregate may be seen (Saxén & Wartiovaara, 1966). L, site of lumen.

Unsworth (1972). At the cellular level, however, further differentiation of the nephric tubules can be demonstrated in prolonged cultures of the mesenchymes in contact with the inductor.

Morphometric measurements have demonstrated that the relative amount of differentiated, epithelial tissue increases linearly until day 6 (Koskimies & Saxén, 1966; Fig. 4.6). Since a maximum response in terms of tubule number is achieved after only 24 h of transfilter culture, this relative increase in the amount of epithelium is due not only to multiplication of the programmed cells but also to an artificial loss of uninduced, less-viable mesenchymal cells.

The basement membrane

From early embryonic stages onwards, many cell types attach to their basement membrane, and this close interaction between a cell and its extracellular substrate seems to be of major importance for cytodifferentiation, cell polarization and multiplication of cells (for reviews, see Wessells, 1977; Gospodarowicz & Tauber, 1980; Bernfield *et al.*, 1984*a*;

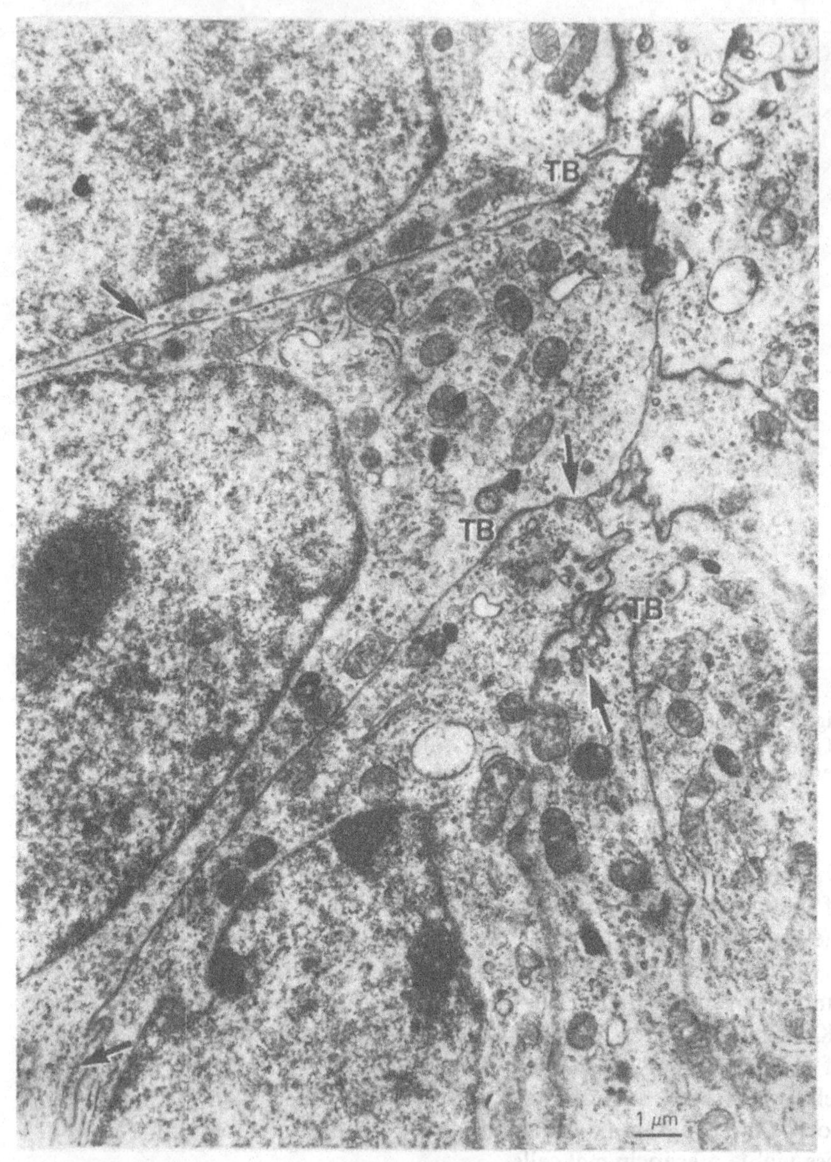

Fig. 4.3. Definite orientation and beginning of lumen formation in a pretubular aggregate. A ring of terminal bars (TB) may be seen in the apical part of the epithelial cells, and numerous Golgi-derived vesicles fuse into the apical cell membrane (arrows) (Saxén & Wartiovaara, 1966).

Ekblom, 1984; Gospodarowicz *et al.*, 1984; Hay, 1984). Similarly, in the embryonic metanephros, clear spatial and temporal associations have been observed between the formation of the basement membrane and differentiation, polarization and tubular assembly of the cells. The

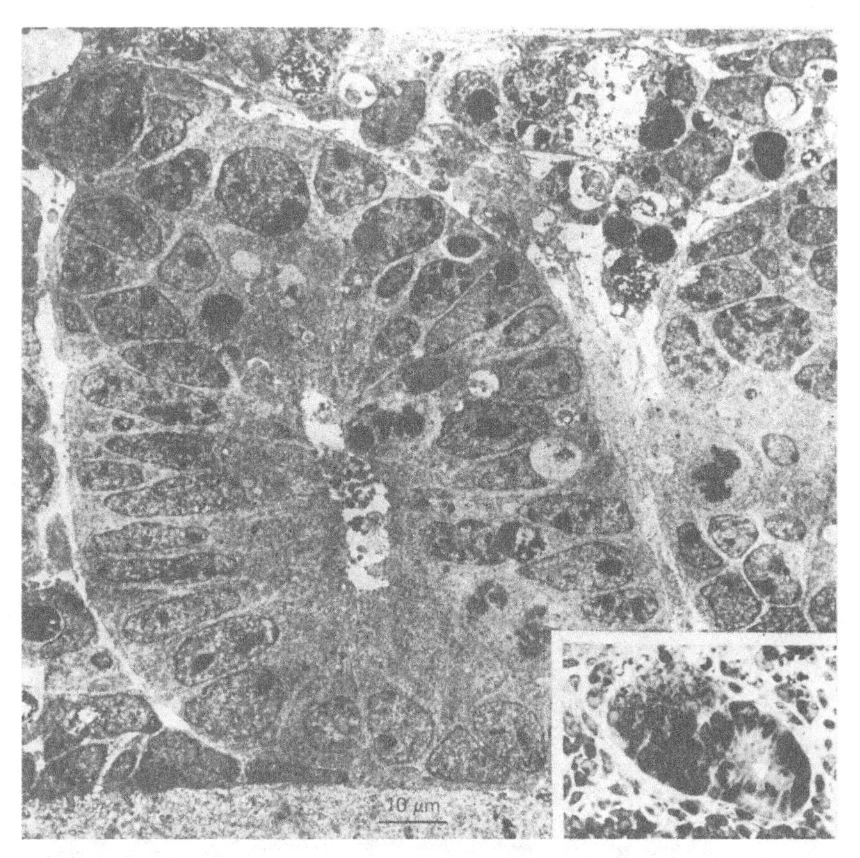

Fig. 4.4. A pretubular renal vesicle in a transfilter-induced mesenchyme with a central lumen and elongated, polarized cells. The inset shows a primitive 'S-shaped' format of the vesicle (Saxén & Wartiovaara, 1966).

model-system for induction and epithelial transformation of the mesenchymal cells *in vitro* seems to offer a unique opportunity to explore this association and the role of the basement membrane in early organogenesis.

Electron microscopy reveals the first signs of the formation of a basement membrane in the induced mesenchymes during the early aggregation phase, when the cells show an onset of polarization (Wartiovaara, 1966*a*; Fig. 4.7). Changes in the endoplasmic reticulum of these cells suggest altered metabolic activity probably associated with the synthesis of basement-membrane compounds. Basement-membrane-like material is then seen to accumulate on the basal surfaces of the elongated cells of the aggregate, and it soon constitutes a continuous basement membrane around the pretubular aggregates/vesicles. No such changes were detected in uninduced mesenchymes or around cells not

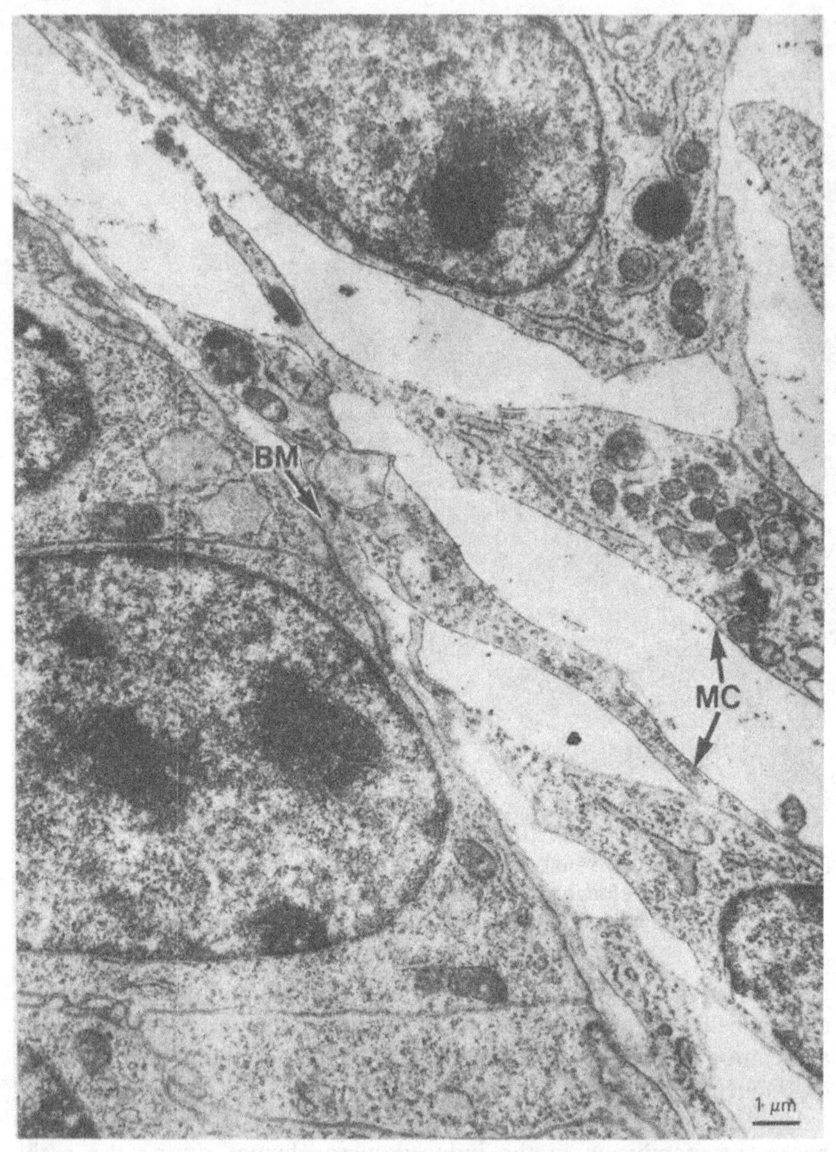

Fig. 4.5. A peripheral portion of a renal vesicle at the stage illustrated in Fig. 4.4. Membrane protrusions have disappeared and basement-membrane-like material (BM) is accumulating on the cell surface but not around the unaggregated mesenchymal cells (MC) (Saxén & Wartiovaara, 1966).

participating in the formation of the aggregates (Wartiovaara, 1966*a*).

Increasing knowledge of the chemical composition of the basement membranes has allowed us to follow these events with specific probes. Many proteins have been found recently in the basement membranes and

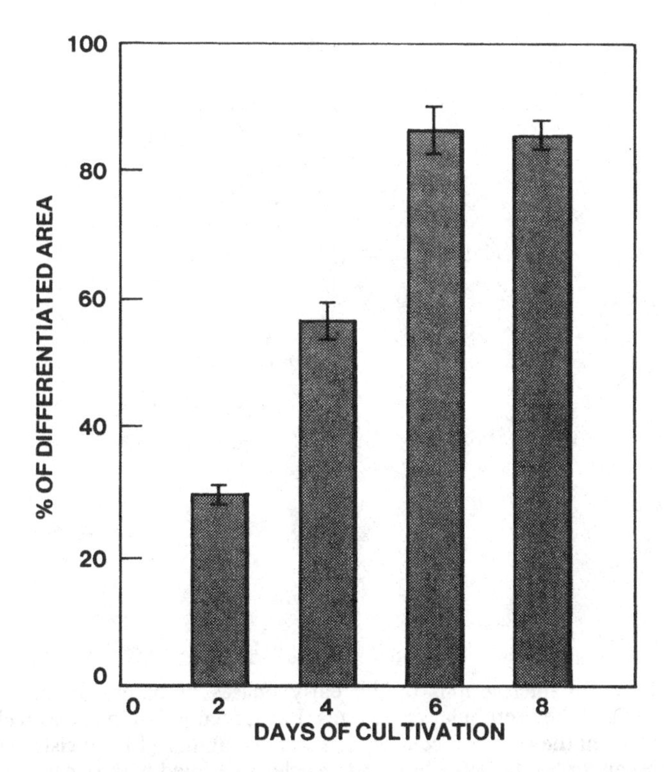

Fig. 4.6. Histogram showing the results of a morphometric measurement of the relative portion of differentiated area in a transfilter-induced mesenchyme (after Koskimies & Saxén, 1966).

characterized in detail (for reviews, see Kefalides *et al.*, 1979; Timpl & Martin, 1982; Hay, 1983, 1984). In addition to collagen types IV and V, the following non-collagenous glycoproteins have been reported to be associated with the basement membrane: laminin (Timpl *et al.*, 1979), heparan sulphate proteoglycan BM–I (Hassell *et al.*, 1980), entactin (Carlin *et al.*, 1981), and nidogen (Timpl *et al.*, 1983). Availability of antisera against these components has made it possible to follow their appearance and distribution in the kidney model-system as well.

Laminin synthesis seems to be an inherent feature of the mesenchymal cells prior to induction. In immunohistology, a weak reaction is observed with an anti-laminin antibody in the 11-day kidney throughout the mesenchyme. Moreover, when these cells are cultivated as monolayers, they synthesize, and deposit, laminin (Saxén & Lehtonen, 1986; Lehtonen & Saxén, 1986*b*; Lehtonen *et al.*, unpublished results). Following induction, however, definite changes in the expression and localization of laminin are detected (Fig. 4.8): distinct fluorescence droplets first appear after some 12 to 24 h *in vitro* and soon begin to disappear from the

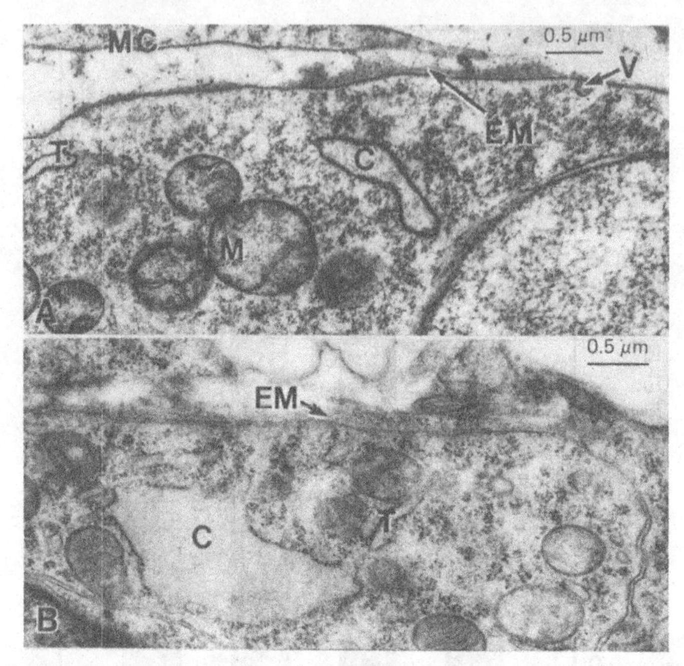

Fig. 4.7. Electron micrographs of the early phases of basement membrane formation around the pretubular aggregates. Local accumulation of extracellular material (EM) on the cell surface is seen as well as tubular (T) and cisternal (C) forms of the endoplasmic reticulum and vesicles (V) filled with dense material (Wartiovaara, 1966*a*; courtesy of Dr J. Wartiovaara). MC, stromal mesenchymal cell; M, mitochondrion.

condensed area. First, they assemble into short linear structures, and then they form a continuous ring around the pretubular aggregates (Ekblom *et al.*, 1980*a*). A similar sequence has been observed for collagen type IV (Ekblom, 1981*a*), heparan sulphate proteoglycan (Lash *et al.*, 1983; Fig. 4.9), and collagen type V (Bonadio *et al.*, 1984).

The proteoglycan synthesis of transfilter-induced mesenchymes has also been analysed in guanidium-extracted samples by Lash *et al.* (1983). In response to induction, a large proteoglycan with a molecular weight of 1 000 000 appeared. The major glycosaminoglycans synthesized were chondroitin sulphates, while heparan sulphate proteoglycans comprised 20% of the proteoglycan fraction. Indirect evidence for the morphogenetic significance of these compounds was obtained in inhibition experiments: 6-diazo-5-oxo-norleucine reduced the glucosamine synthesis by 60% in a concentration previously reported to prevent tubule formation (Ekblom *et al.*, 1979*a*; Lash *et al.*, 1983).

In conclusion, both ultrastructural and immunohistological as well as biochemical data show that, during early aggregation of the induced

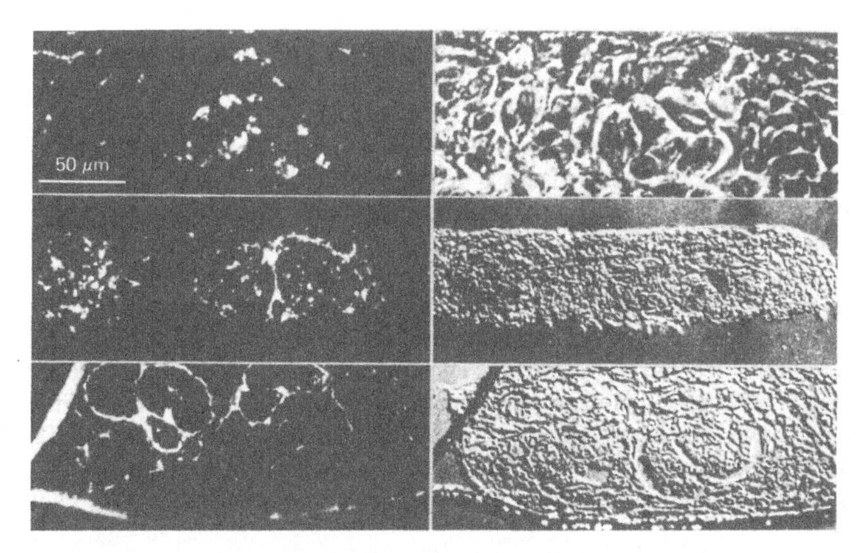

Fig. 4.8. Fluorescence micrographs of different phases of laminin expression in metanephrogenic mesenchymes in transfilter contact with an inductor. Hours of contact are 36, 48 and 72, from top to bottom, respectively. The initially random distribution of the fluorescence droplets gradually adopts a linear appearance, ultimately becoming a continuous circle surrounding the tubules. *Left:* sections treated with anti-laminin antibody. *Right:* phase-contrast micrographs of the same sections (Ekblom *et al.*, 1980*a*).

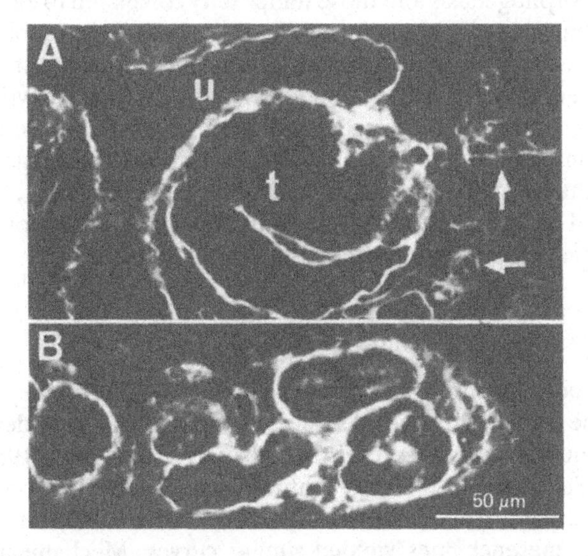

Fig. 4.9. Fluorescence micrographs demonstrating the distribution of heparan sulphate proteoglycan in a section from a 12-day whole-kidney rudiment (A) and from a transfilter culture of three days (B). Treated with antiserum against the basement-membrane proteoglycan BM-1 (Lash *et al.*, 1983). u, ureter; t, tubule; arrows, capillaries.

mesenchymal cell, constituents of the basement membrane are synthesized, and this leads to a complete basement membrane at the renal vesicle stage. The data do not necessarily imply that the basement membrane is the polarizing factor – rather it seems that its basal location is a consequence of polarization of the cells within an aggregate. An increased adhesion of the induced cells packs them into compact condensates where polarization might be brought about by lateral adhesive properties, as suggested originally by Gustafsson & Wolpert (1963). This polarization leads to apical lumen formation and basal accumulation of the basement membrane components. Subsequently, the basement membrane becomes the anchoring substrate of the cells and might have a central role in the maintenance of the polarized state. This view is well in accordance with our observations (p. 122) that induced cells in monolayer cultures express epithelial differentiation but remain unpolarized with random localization of the basement membrane components.

Metabolic changes following induction

Introduction

The transfilter model-system for kidney development is also well suited for analysis of metabolic changes in the target mesenchyme. Changes preceding morphogenesis and those temporally correlated to early aggregation and subsequent tubule formation have both been reported. The apparent drawback of many such studies is the tiny amount of tissue available. Hence, systemic fractionation experiments and analytical work have only recently been made feasible with advanced, more sensitive techniques in biochemistry and immunochemistry. Knowledge of early metabolic changes following an inductive trigger is, however, still fragmentary and may not lend itself to comprehensive conclusions on the molecular background of induction and morphogenesis.

Early studies of metabolic changes

Early metabolic changes in the experimentally induced nephrogenic mesenchyme were first reported some 20 years ago and were detected at the level of nucleic acid and protein synthesis by Vainio *et al.* (1965) (for a review see Saxén *et al.*, 1968). Quantitative studies on incorporation of radiolabelled precursors (uridine, thymidine and leucine) into isolated metanephric mesenchymes yielded similar curves. Mechanical separation of the tissues and their transfer to culture conditions resulted in a metabolic 'standstill', shown as a gradually decreasing incorporation rate of the labelled compounds. In the culture conditions then applied, this adaptation phase lasted for 16 to 20 h from the establishment of the transfilter contact with the spinal cord. After this, the incorporation rate rose rapidly in the induced mesenchymes. A clear but less striking recovery was also observed in the uninduced mesenchymal explants.

Fractionation results related to the RNAs synthesized after the lag period might still be of interest, though they were obtained long ago and now call for the re-examination by modern technology. Maximum incorporation of [^{3}H]uridine was measured 30 h after setting up the culture (Vainio *et al.*, 1965). It was then five times that of the uninduced mesenchymes in parallel cultures. The greatest difference was noted in the rRNA fraction (Miettinen *et al.*, 1966). At the cellular level, the cytoplasmic RNA concentration was measured by ultraviolet microspectrophotometry (Sundelin *et al.*, 1969). At 24 h after setting up the transfilter culture, cells in the induced mesenchymes showed an elevated cytoplasmic RNA as compared with uninduced controls. At the early condensation stage (total culture time 36 h) cells within the aggregates and pretubular renal vesicles showed somewhat higher cytoplasmic RNA than did cells remaining in the uncondensed stroma.

All the above observations show that there is a clear enhancement of synthesis of RNA in the induced mesenchymal cells towards the end of the induction period and soon thereafter. Apart from the temporal correlation between this increased synthesis and the determination of the mesenchyme, further emphasis on the morphogenetic significance of the increased synthesis is obtained from experiments which made use of actinomycin D, an inhibitor of RNA synthesis. When applied prior to completion of induction, i.e. during the first 24 h of transfilter contact, the drug completely prevented subsequent morphogenesis (Jainchill *et al.*, 1964; Koskimies, 1967*a*). After this sensitive period, the mesenchyme is resistant to actinomycin D in non-lethal concentrations. Further conclusions on the role of increased RNA synthesis and the effect of actinomycin D are hampered by incomplete knowledge of the action of the drug and by difficulties in eliminating its directly toxic, non-specific effects. The isolated mesenchyme, transferred *in vitro*, has proved extremely sensitive to various chemicals in the culture medium, and all conclusions on any inhibitory actions should be made with caution.

From the results indicating increased RNA synthesis as a response to induction, one would naturally predict enhanced protein synthesis as well, and this was also suggested by the increased incorporation of radioactive leucine towards the end of the induction period (Vainio *et al.*, 1965). Appearance of new proteins during and soon after this period has been suggested by many observations, but the precise time is naturally greatly affected by the sensitivity of the detection methods. According to Rapola & Niemi (1965), early condensation of the mesenchymal cells is accompanied by an increased activity of the enzyme nicotinamide adenine dinucleotide-tetrazolium reductase, and somewhat later acid

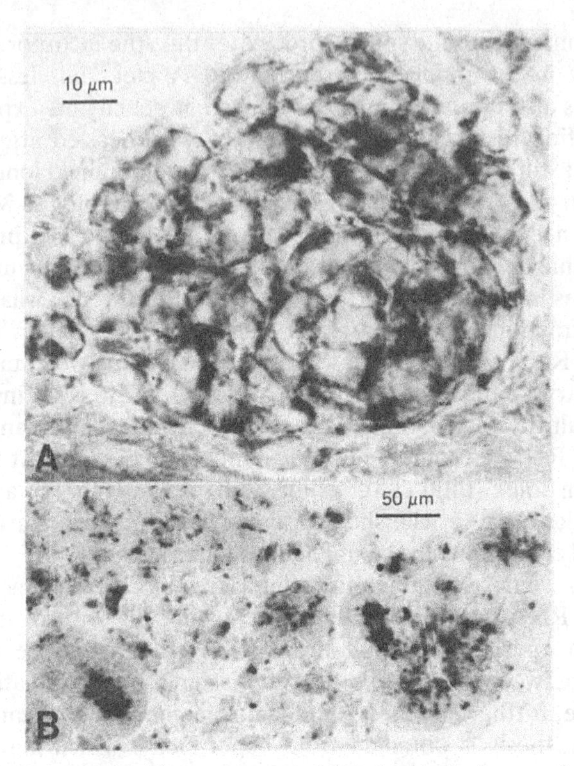

Fig. 4.10. Micrographs illustrating two histochemical changes in an experimentally induced metanephric mesenchyme (Rapola & Niemi, 1965). A. Nicotinamide adenine dinucleotide-tetrazolium reductase activity in a mesenchymal cell condensate. B. Acid phosphatase activity at the luminal portion of the cells of an advanced tubule on day 5 of cultivation.

phosphatase activity can be detected in the apical cytoplasm of the polarized epithelial cells (Fig. 4.10).

The enzyme lactate dehydrogenase (LDH) occurs in five tetramers composed of two polypeptide subunits (A and B) which combine in all possible combinations (LDH 1 = A^0B^4, LDH 2 = A^1B^3 and so on; Markert & Ursprung, 1962). The undifferentiated metanephric mesenchyme exhibits an embryonic-type LDH pattern with the A subunit predominating. After induction, the ratio A:B changes towards the polypeptide B and at 48 h a clear shift towards the anodal LDH bands on polyacrylamide can be detected. On the tenth day of cultivation *in vitro* the LDH pattern closely resembles that observed in an adult mouse kidney (Koskimies & Saxén, 1966; Fig. 4.11). The apparent shift in the expression of the two polypeptides could also be inhibited by actinomycin D (Koskimies, 1967*b*). Interestingly, the effect of this drug on morphogenesis (p. 67) could be temporally dissected from the effect on the LDH shift (the ratio of subunit A to subunit B). Tubule formation was

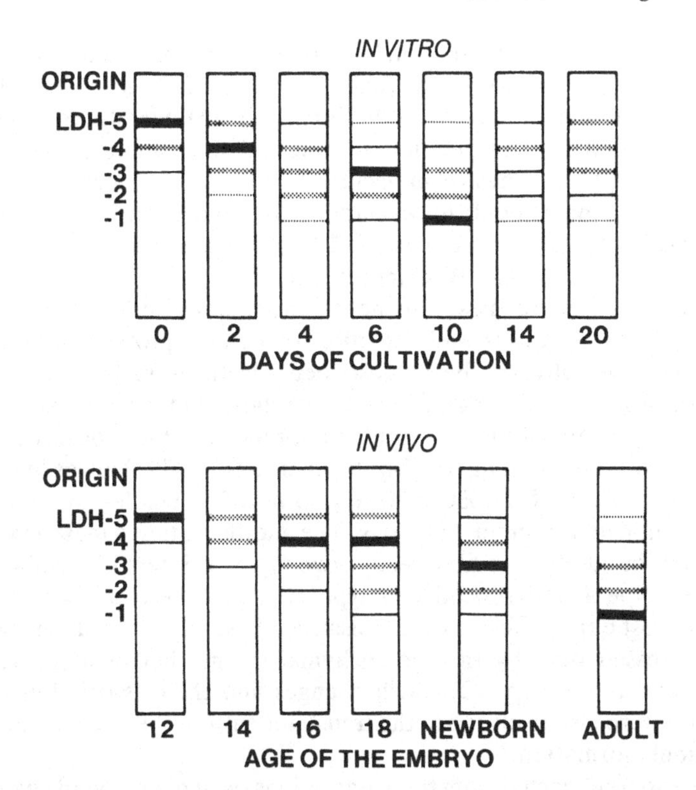

Fig. 4.11. Scheme of electrophoretograms demonstrating the shift in the lactate dehydrogenase (LDH) isozyme pattern in transfilter-induced mesenchyme (upper) as compared with changes *in vivo* during organogenesis (after Koskimies, 1967*b*).

prevented by treatment up to 24 h, whereas the shift in the LDH pattern could still be inhibited by identical treatment up to 30 h.

Changes in the extracellular matrix

There are several reasons for predicting that postinductory changes may occur in the composition of the extracellular matrix (ECM) of the nephric mesenchyme. Such matrices are synthesized by most embryonic and adult cell types and laid down between the cells, where they create many of the specific morphological and functional features of the different tissues. Profound changes in such characteristics during embryogenesis could probably be associated with changes in the composition of the ECM. As suggested by Grobstein (1955*b*), the components of the ECM might be involved also in the morphogenetic tissue interactions between cells or between cells and their matrices. An interaction between cells and their matrix substrate has also been suggested frequently to be of major importance for cell differentiation, locomotion and proliferation (Lash & Vasan, 1978; Gospodarowicz & Tauber, 1980; Saxén *et al.*, 1982; Hay, 1983, 1984; Bernfield *et al.*, 1984*a*, *b*; Ekblom, 1984). Finally, cell-to-cell attachment, vital for the formation of tissue and organ anlagen, is known to be mediated by adhesive molecules at the cell surfaces or in the extracellular compartment (for example, Moscona, 1974; Burger, 1974; Wartiovaara *et al.*, 1980; Edelman, 1983, 1985; Thiery *et al.*, 1984; Yoshida-Noro *et al.*, 1984; Birchmeier *et al.*, 1985).

In the metanephric mesenchyme experimentally induced *in vitro*, changes in the synthesis and secretion of certain proteins following induction have already been described. Both ultrastructural and immunohistological findings showed secretion and accumulation of certain components of the basement membrane, and these observations correlated well with findings *in vivo*. Furthermore, early changes in some other components of the ECM were also noted in whole-kidney rudiments examined by immunohistology. The undifferentiated mesenchyme expressed, in addition to fibronectin, two types of interstitial collagens (collagen type I and procollagen type III). All these proteins were gradually lost during the early condensation phase *in vivo*, and they were then expressed only by the mesenchymal stroma between the renal vesicles and tubules (p. 22). Such changes and their association with induction were also explored in the transfilter cultures before and during early tubule formation.

As was to be expected from the observations *in vivo*, profound changes in the composition of the mesenchymal ECM were recorded after transfilter induction (Ekblom *et al.*, 1981*b*; Fig. 4.12): an uninduced mesenchyme, grafted without any inductor on to the filter, uniformly expressed fibronectin, collagen type I and procollagen type III when explored by immunomicroscopy. In contrast, when the mesenchyme was combined transfilter with the inductor (spinal cord), the three ECM components seemed to be lost rapidly; after 6 h of cultivation, a negative, basal zone was seen nearest to the filter surface, and after an additional 6 h only the upper third of the mesenchyme expressed the three components of the ECM. After a total of 24 h in contact with the inductor, all three proteins were lost from the mesenchyme, and only traces of them were detected on its surface. Temporal correlation of these changes with the onset and completion of induction in the same conditions is evident.

In order further to correlate the loss of the interstitial proteins with the induction of the nephric mesenchyme, previous findings on the 'specificity' of the inductive interactions in kidney development were utilized. By purely morphological criteria, it has been shown that the nephric mesenchyme is the only embryonic tissue responding to an inductor by formation of tubules, and that many heterologous tissues exert this action mimicking normal induction (p. 66). Consequently, to study the possible changes in the ECM, reciprocal combinations were prepared between

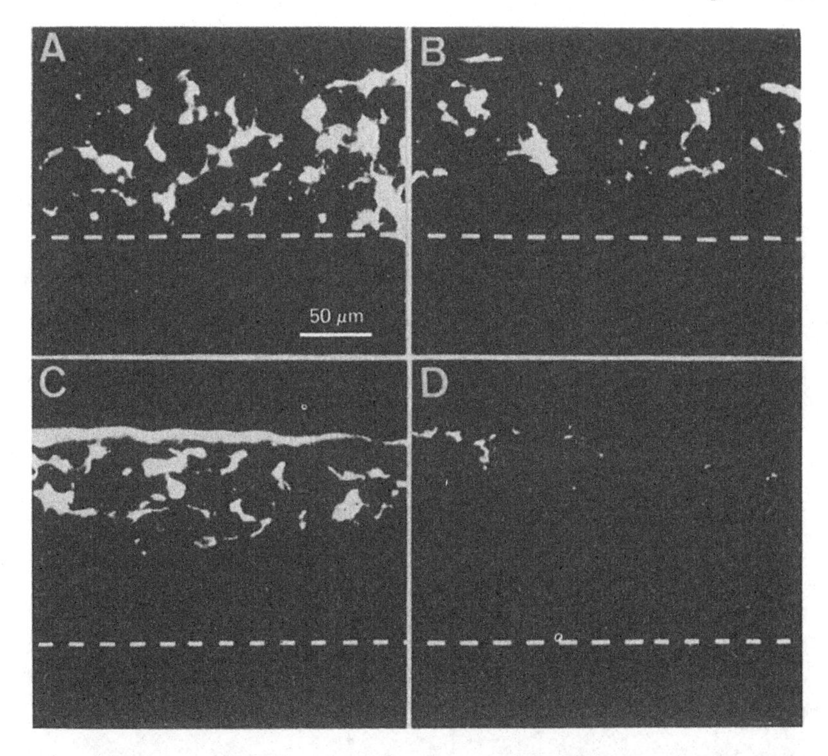

Fig. 4.12. Series of fluorescence micrographs illustrating the loss of collagen type III from transfilter-induced mesenchymes. Antibodies against procollagen type III applied to frozen sections of mesenchymes harvested at different intervals after the setting up of a transfilter explant. The dashed white line indicates the upper surface of the filter (not seen in the micrographs) (Ekblom *et al.*, 1981b). A. Mesenchyme cultivated for 6 h without underlying inductor. B. Mesenchyme cultivated for 6 h in transfilter contact with the inductor. Note the fluorescent-negative zone against the filter. C. Mesenchyme cultivated for 12 h in contact with the inductor. D. Mesenchyme cultivated for 24 h in contact with the inductor. Only traces of antigenicity can be detected at the surface of the explant.

non-nephrogenic mesenchymes and tubule inductors on the one hand and between the nephrogenic mesenchyme and non-inducing tissues on the other. Such combinations never led to the above-mentioned changes in the ECM of the mesenchyme, whereas a technically identical recombinant of the ureteric bud and the nephric mesenchyme reproduced the earlier findings (Ekblom *et al.*, 1981*b*; Fig. 4.13.).

In conclusion, induction of the metanephric mesenchyme can in several ways be associated with a rapid loss of certain interstitial proteins from the target tissue: (1) *in vivo*, a close spatial association was demonstrated between the induced, aggregating zone around the ureter and the loss of the three components of the ECM; (2) in the transfilter cultures this loss showed a close temporal correlation with the onset and completion of induction; and, finally (3), changes in the composition of

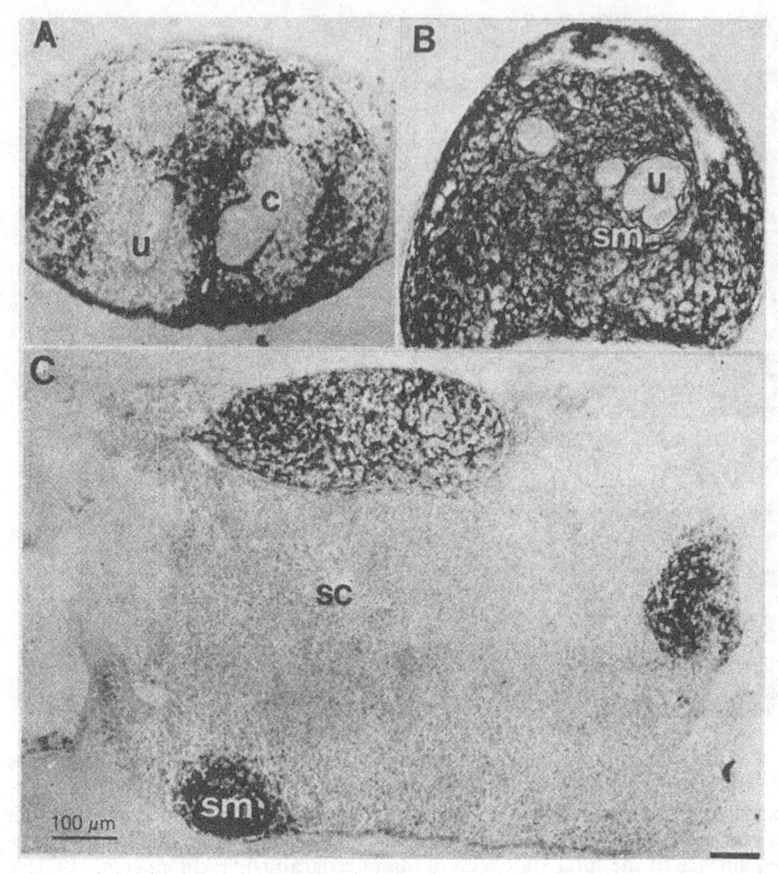

Fig. 4.13. Micrographs demonstrating the induction-dependent loss of collagen type III from nephrogenic mesenchyme. Sections treated with an antiserum against procollagen type III and stained by the peroxidase technique (Ekblom *et al.*, 1981*b*). A. An experimental recombination of isolated metanephric mesenchyme to fragments of the ureteric bud. Note the negative halos surrounding the ureter (u). B. A recombinant (as in A) where the ureteric tissue (u) has been combined with non-nephric (salivary gland) mesenchyme. No loss of collagen is seen. C. An experimental combination of an inductor (spinal cord, sc) to three fragments of embryonic salivary gland mesenchyme (sm). No changes in the expression of procollagen type III are detectable.

the ECM could not be detected in non-inductive situations. The possible causal relations between induction, early aggregation, and the changes in the interstitial proteins will be discussed in Chapter 6.

Changes in the cytoskeleton

The type, distribution and assembly of the intracellular filaments constituting the cytoskeleton are specific features of tissues and cell lineages

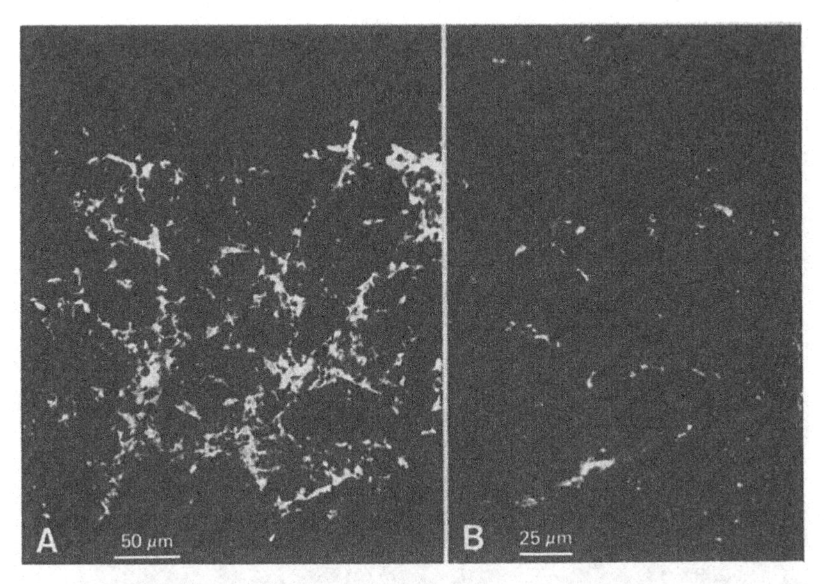

Fig. 4.14. Fluorescence micrographs demonstrating the gradual restriction of cells expressing vimentin-type intermediate filaments in an experimentally induced mesenchyme. Sections were treated with anti-vimentin antibody (Lehtonen *et al.*, 1985; courtesy of Dr E. Lehtonen). A. Section of a mesenchyme cultivated for four days in transfilter contact with spinal cord. B. A section through a mesenchyme harvested on day 7 of transfilter cultivation.

which change from cell to cell and during embryogenesis (Franke *et al.*, 1982; Holtzer *et al.*, 1982; Lazarides, 1982). Their roles in morphogenesis, cell polarization and cytodifferentiation are not yet fully understood, but an interaction between certain 'intermediate size' filaments and the extracellular matrix has been suggested as a morphogenetically significant mechanism (Hay, 1984).

I have already described some morphogenesis-associated changes in the distribution and composition of the intermediate filaments in cells of human and murine embryonic kidneys *in vivo*. Only preliminary data are available for similar changes in the experimentally induced mesenchymal cells which are being converted into epithelial cells and structures. The undifferentiated, uninduced nephric mesenchyme expresses vimentin-type intermediate filaments. No apparent changes are observed during cultivation without an inductor. When induced by the transfilter technique, cells that become aggregated into pretubular condensates lose their vimentin filaments (around day 4), and during prolonged cultivation vimentin becomes confined to the narrow, intertubular strands of undifferentiated stroma (Lehtonen *et al.*, 1985; Fig. 4.14). A reverse mode of development may be observed when the mesenchymes are examined with several anti-cytokeratin antibodies as probes: the undifferentiated

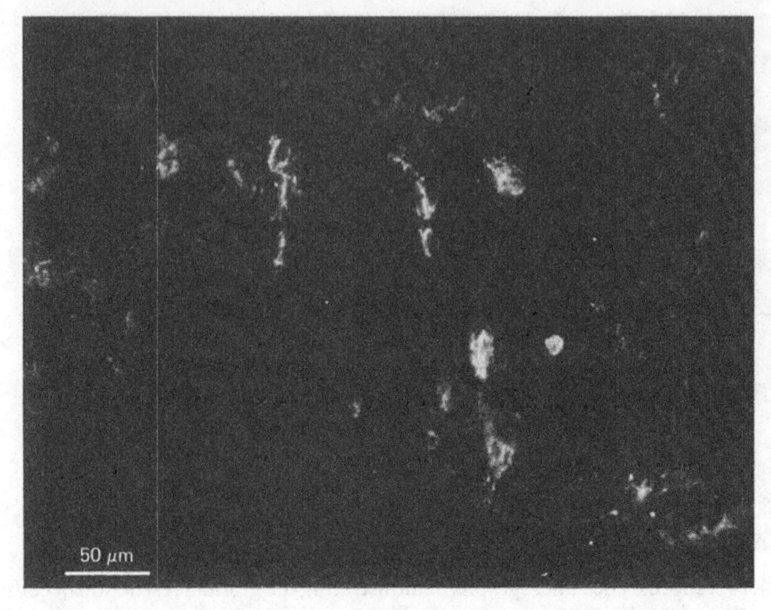

Fig. 4.15. Fluorescence micrograph demonstrating the expression of cytokeratin-type filaments in a mesenchyme cultivated for 7 days in transfilter contact with an inductor (Lehtonen *et al.*, 1985; courtesy of Dr E. Lehtonen).

mesenchymal blastema is devoid of any signs of cytokeratin, which appears on day 4 as traces in cells within the condensates. The number of cells expressing cytokeratin-type intermediate filaments increases gradually, and on day 7 most of the epithelial cells of the tubules are positive when examined by immunohistology (Lehtonen *et al.*, 1985; Fig. 4.15.).

Proliferation

Introduction

One early response to an inductive stimulus might be a change in the proliferative state of the target tissue. In fact, this has been shown in the epithelial component of many glandular organs and of the integument after an interaction with the mesenchymal stroma (Alescio & Cassini, 1962; Wessells, 1963, 1970; Spooner & Hilfer, 1971; Ronzio & Rutter, 1973; Osman & Ruch, 1978). In the kidney model-system, both direct measurements and indirect experimental data have provided evidence for an association between induction of the metanephric mesenchyme and changes in its DNA synthesis. Vainio *et al.* (1965) observed an increased uptake of [^{3}H]thymidine by the nephric mesenchyme after 30-h transfilter contact with an inductor. After this peak of maximal incorporation, thymidine uptake diminished almost to the level of the controls, i.e. mesenchymes cultivated in identical conditions without exposure to an inductor.

The suggestion of a morphogenetic role for the postinductory enhancement of DNA synthesis is compatible with findings that an inhibition of this synthesis during or soon after the 'induction period' prevents subsequent epithelial transformation and tubule formation of the exposed mesenchyme. Interestingly, Sobel (1966) reported that this inhibition of morphogenesis is to a certain extent reversible: if the tissues treated with 5-fluorodeoxy-uridine during the first 20 h of contact were released from the effect of the inhibitor, disorganized tubules formed after a lag period. Nordling *et al.* (1978) used mitomycin C, an inhibitor of DNA synthesis, during the initial stages of induction and found a dose-dependent inhibition of tubule formation. The active concentration of the drug correlated well with the measured inhibition of thymidine incorporation (Fig. 3.7, p. 68). But conclusions were once again hampered by the fact that, like many other metabolic inhibitors, mitomycin C may interfere with the synthesis of macromolecules other than DNA.

Stimulation of DNA synthesis

More recently we have performed a detailed study of changes in the incorporation of thymidine into the metanephrogenic mesenchyme after transfer and induction *in vitro* (Saxén *et al.*, 1983). Dissected mesenchymes were combined routinely with a fragment of inducing spinal cord through interposed filter membranes, and incorporation of [^{3}H]thymidine was recorded after different time intervals. Incorporation, calculated per total DNA of the mesenchymes, was compared with uninduced controls, otherwise treated and pulse-labelled in an identical way. The results (Figs. 4.16 and 4.17) show initial depression of thymidine incorporation during the first 10 h of cultivation. This was evident both in mesenchymes cultivated in transfilter contact with the inductor and, to a lesser extent, in those cultured in isolation. After the period of low activity, the induced mesenchyme entered a phase of high incorporation that reached its peak around 24 h and then declined. The uptake of thymidine remained, however, two to three times higher in the induced mesenchymes than in the uninduced ones.

Superimposed on the earlier data on induction and response of the nephric mesenchyme, the incorporation data show a good temporal correlation (Fig. 4.17). The initial phase of low metabolic activity up to 10 h correlates with the period when induction has not yet occurred. Induction takes place during the following 12- to 24-h period, and more cells gradually enter the programme of epithelial differentiation. At the end of this period, maximal incorporation was measured and the time

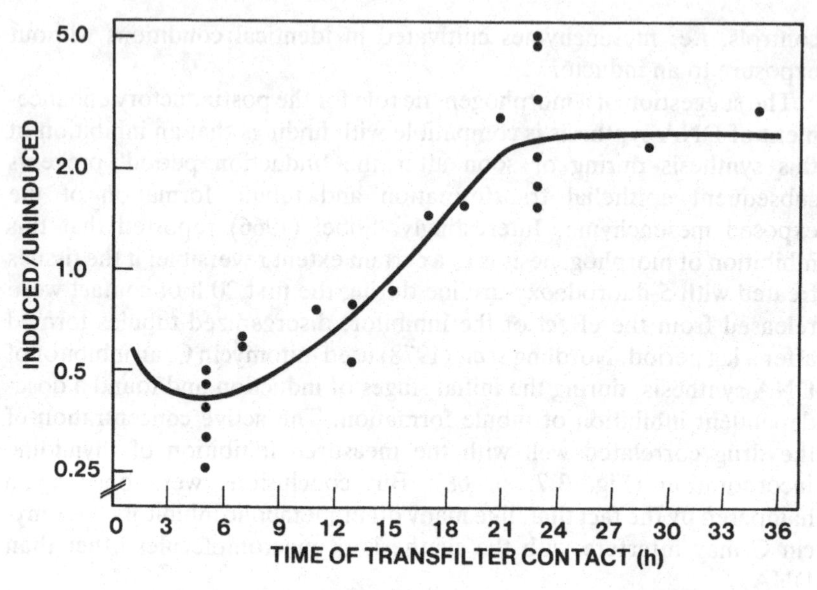

Fig. 4.16. Incorporation of tritiated thymidine into metanephrogenic mesenchyme *in vitro*. Uptake recorded as [^{3}H]thymidine incorporation, c.p.m./μg of DNA and given as a ratio of induced/uninduced mesenchyme (after Saxén *et al.*, 1983).

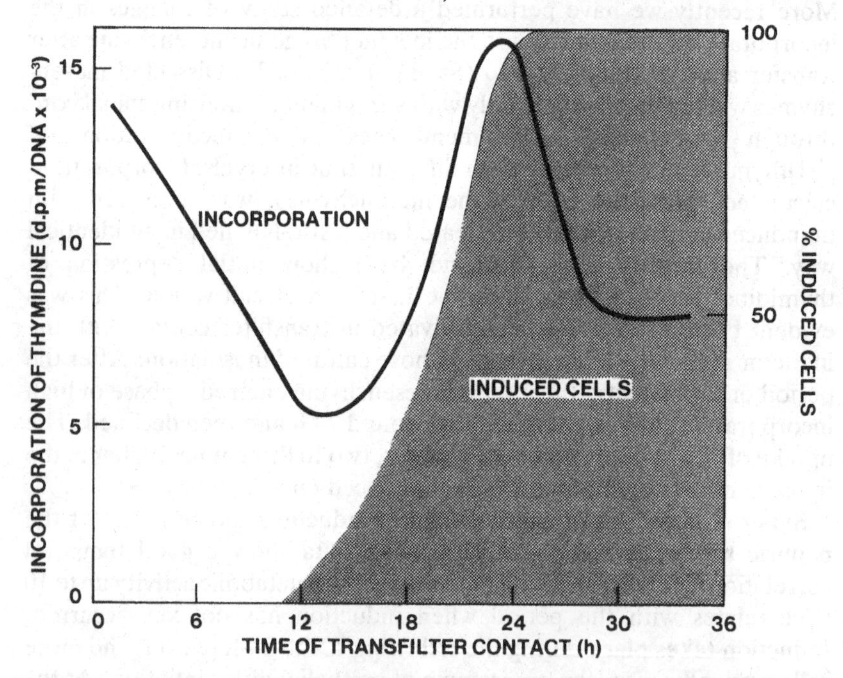

Fig. 4.17. Incorporation curve of [^{3}H]thymidine into transfilter-induced mesenchyme superimposed onto the 'induction curve' from Fig. 3.18 (hatched area) (after Saxén & Lehtonen, 1978; Saxén *et al.*, 1983).

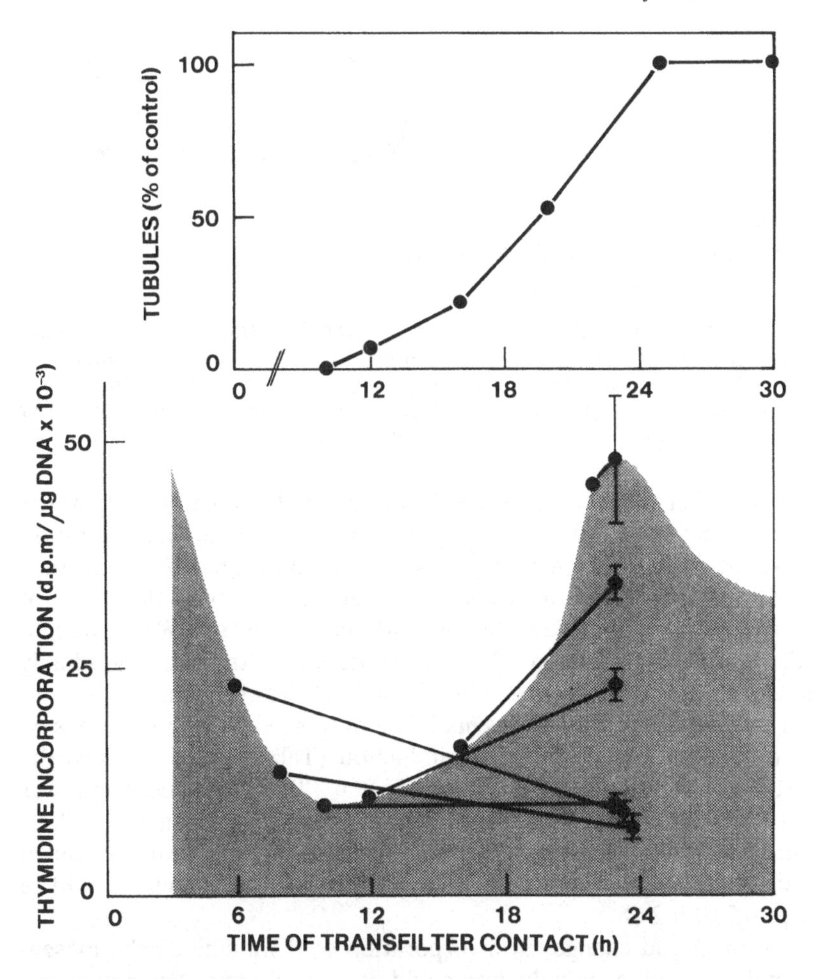

Fig. 4.18. The effect of an interrupted induction on the formation of tubules (above) and on the incorporation of [^{3}H]thymidine into nephric mesenchymes in a transfilter culture (below). The inductor was removed at different durations and different phases of incorporation (stippled), and incorporation of [^{3}H]thymidine was measured at 24 h. The connecting lines thus indicate change in incorporation after interruption of induction (after Saxén *et al.*, 1983).

corresponds to the completion of induction, when all cells to be determined have entered the programme. After this maximum, thymidine incorporation slowly drops to a lower level and the time corresponds to the stage when the early events of cytodifferentiation become detectable in the induced mesenchyme.

Two further experiments link the changes in DNA synthesis to the effect of the inductor. In one of these, thymidine incorporation was followed after interrupted induction (Fig. 4.18): the inductor was

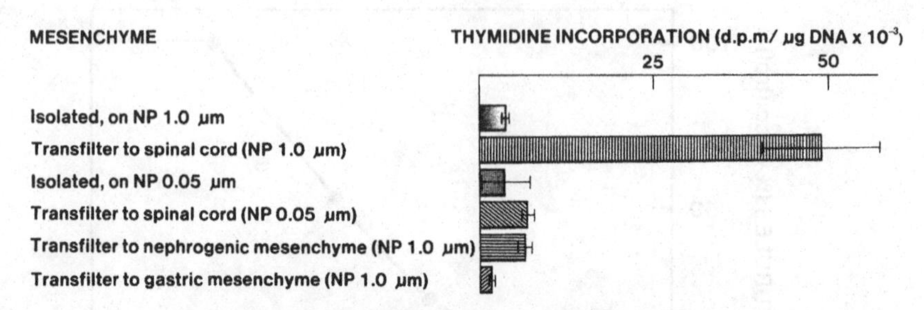

Fig. 4.19. Incorporation of [^{3}H]thymidine after 24-h transfilter contact with inducing and non-inducing tissues after cultivation in isolation or combined with an inductor through a Nuclepore (NP) filter preventing induction. Only conditions resulting in subsequent tubule formation stimulate incorporation (after Saxén *et al.*, 1983, 1985).

removed after different intervals during the 24-h induction period, the mesenchyme was subcultivated and its thymidine incorporation measured at 24 h. Mesenchymes in which the contact had been interrupted during the first 10 h of cultivation did not differ in their rate of incorporation from those cultivated without an inductor. When the first cells had been induced at 12 h, incorporation of thymidine gradually increased, even after contact with the inductor was broken.

A second set of experiments made use of some earlier observations on the transfilter passage of tubule induction (Table 3.4, p. 75) and the existence of non-inducing tissues (Fig. 4.19). When induction was prevented by interposing filters with pore-size less than the critical diameter or when the mesenchymes were combined with tissues devoid of inductive activity, no increase in the incorporation of thymidine could be detected after 24 h.

The profound changes in incorporation of thymidine by the mesenchymal cells following induction could be due to one of two responses: either the number of cells in S-phase or the length of the entire cell cycle varies during the different periods. Direct flow cytometric counts of cells at different phases of the cycle seemed to rule out the first alternative, and it was concluded that the probable mode of response to induction is a shortened cell cycle (Saxén *et al.*, 1983). It still remains unsettled whether this inductive stimulus is 'directive', causing both determination and proliferation on the mesenchymal cells, or merely 'permissive', stimulating multiplication of a predetermined cell population within the mesenchymal blastema. Such a stimulation could be implemented by production and secretion of a mitogen by the inductor or by rendering the mesenchymal target cells responsive to soluble growth factors. Recent experimental results suggest that both mechanisms are involved.

Growth factors

All the above experiments recording the proliferative response of the nephric mesenchyme to an inductive stimulus were performed *in vitro* by using culture media supplemented with serum. Serum, on the other hand, might contain hormones (e.g. insulin, somatomedins) and low molecular weight 'growth factors' (e.g. epidermal growth factor, nerve growth factor, platelet-derived growth factor), all exhibiting a definite mitogenic effect on various cell types (for reviews, see Barnes & Sato, 1980; Gospodarowicz & Tauber, 1980). Tests *in vitro* for exploring such factors follow two strategies: cells and tissues can be cultivated in sera selectively depleted of such compounds; or cells can be cultured in defined media devoid of unknown protein constituents, and the factors to be examined can be added to these. The experimental situation, however, still remains complex, as many compounds with the mitogenic effect can interact with other constituents of the culture media or with the substrata to which cells attach.

After some preliminary data suggested that kidney tubule morphogenesis can be initiated in a chemically defined, serum-free medium to which transferrin, an iron chelator, is added (Ekblom *et al.*, 1981*c*), a systemic analysis was performed by Ekblom and collaborators (Ekblom *et al.*, 1983; Thesleff & Ekblom, 1984; Ekblom & Thesleff, 1985). The results convincingly showed the importance of transferrin to the proliferation of the mesenchymal cells and to the morphogenesis of the kidney. However, the first phase of a stimulated incorporation of thymidine between 12 and 24 h seemed to be transferrin independent, as it occurred in the chemically defined medium in the absence of transferrin (Fig. 4.20). After a peak incorporation at 24 h, the induced cells became transferrin dependent. Moreover, parallel cultures of isolated mesenchymes not exposed to an inductor remained fully unresponsive to transferrin. Induction of the mesenchymal cells in the metanephric blastema evidently renders them responsive to transferrin, which subsequently acts in this system as a growth factor.

The above results do not favour the idea that transferrin is released from the inductor and acts upon the induced mesenchymal cells – thus far transferrin has not been detected in the inductor tissue, but has been found in the serum at the time of onset of kidney organogenesis (11-day mouse embryos). Ekblom & Thesleff (1985) found transferrin in the foetal liver and in the visceral yolk sac and showed that the liver stimulated thymidine incorporation into the metanephric mesenchyme, provided that the mesenchyme was in contact with a tubule inductor (Fig. 4.21). The effect of the liver tissue upon the nephric mesenchyme was not contact dependent and could be prevented by anti-transferrin antibodies. Furthermore, the authors showed that transferrin-depleted serum did not

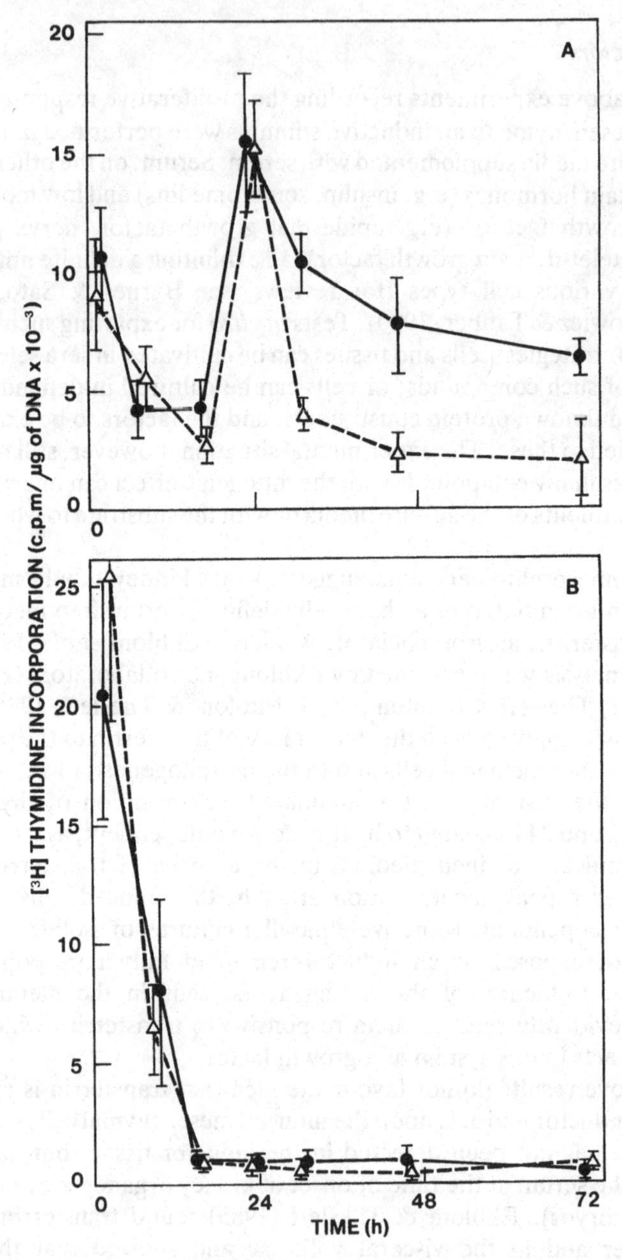

Fig. 4.20. Incorporation of [^{3}H]thymidine into transfilter-induced metanephric mesenchymes (A) and uninduced controls (B). Continuous line (●—●), cultivated in the presence of transferrin at 50 μg/ml; broken line (△——△), no transferrin added to the chemically defined medium (after Ekblom *et al.*, 1983).

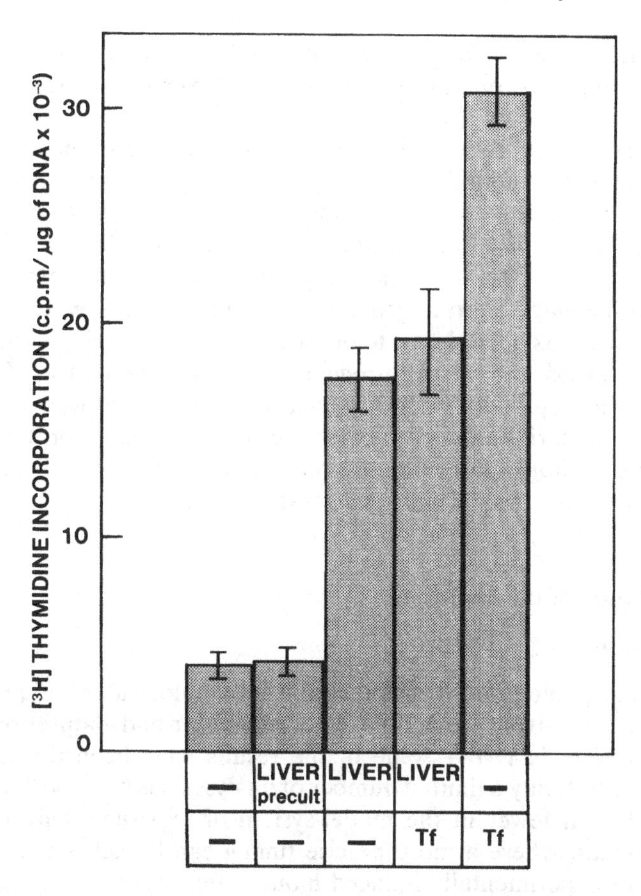

Fig. 4.21. The stimulatory effect of embryonic liver tissue grafted with induced metanephric mesenchyme either in the presence or in the absence of transferrin (Tf). Medium conditioned with liver (precult) had no effect (after Ekblom & Thesleff, 1985).

support growth and morphogenesis of the kidney (Thesleff & Ekblom, 1984).

The mode of action of transferrin on the induced mesenchyme is not yet fully understood, but it seems to be associated with the transmembrane iron transport mediated by transferrin. This has been shown indirectly by Landschulz *et al.* (1984), who used a non-physiological, lipophilic iron chelator, pyridoxalisonicotinoylhydrazone (PIH), to which ferric iron binds to form the complex FePIH. In the absence of either transferrin or FePIH, an excess of ferric iron in the culture medium did not support proliferation of the kidney mesenchymal cells, whereas FePIH mimicked the action of transferrin in apparently by-passing the transferrin-mediated transfer of iron from the exterior to the intracellular compartment. Hence, the authors concluded that the observed effect of transferrin on cell proliferation was 'solely to provide iron' (Landschulz *et al.*, 1984).

The induction of transferrin-responsiveness of the mesenchymal cells during the initial period of contact has not yet been explained. A plausible mechanism would be induction of an increased number of transferrin receptors at the surface of the target cells, but thus far conclusive evidence for this is lacking. Such induction of receptors during a contact-mediated morphogenetic interaction of cells would not be unique, and it has been shown to occur during early development of the mammary gland and in the urogenital tract of the mouse. In both instances, an epithelial–mesenchymal interaction induces hormone receptors, rendering the cells responsive to organismal morphogenetic stimuli, i.e. to hormones with developmental effects (Kratochwil, 1977; Kratochwil *et al.*, 1979; Cunha *et al.*, 1983, 1985).

Segmentation of the nephron

Introduction

Various morphological, biochemical and immunological techniques have been used extensively to examine the segmentation and maturation of the nephric vesicle *in vivo* – some of the results have been described in Chapter 1. But only a limited number of analyses has been performed *in vitro* and even fewer in the model-system of experimentally induced mesenchymes, where a more precise timing can be achieved. In such cultures of experimentally induced mouse metanephric mesenchymes, advanced, coiled tubules and glomerulus-like bodies have been documented by Koskimies (1967*a*) and by Gossens & Unsworth (1972). More recently Bernstein *et al.* (1981) have performed a detailed analysis at the ultrastructural level. They describe well-shaped though avascular glomeruli in metanephric cultures induced transfilter by spinal cord and cultivated for seven days. Staining with colloidal iron visualized a polyanionic coat on the podocyte surface and an arborizing basement membrane of the epithelial cells (Figs. 4.22 and 5.4, p. 138).

Early immunohistological studies have likewise suggested late maturation of such cultures. After Okada & Sato (1963) and Okada (1965) had demonstrated an adult-type microsomal, apparently kidney-specific antigen(s) in the proximal tubules of chick mesonephros, we followed the line in the metanephros of the mouse and in experimentally induced metanephric mesenchymes (Lahti & Saxén, 1966). The appearance of an adult-type antigen was detected in the kidneys of 16-day embryos and on day 12 in the experimentally induced mesenchymes *in vitro* (Fig. 4.23). The observation shows that differentiation of the nephron *in vitro*

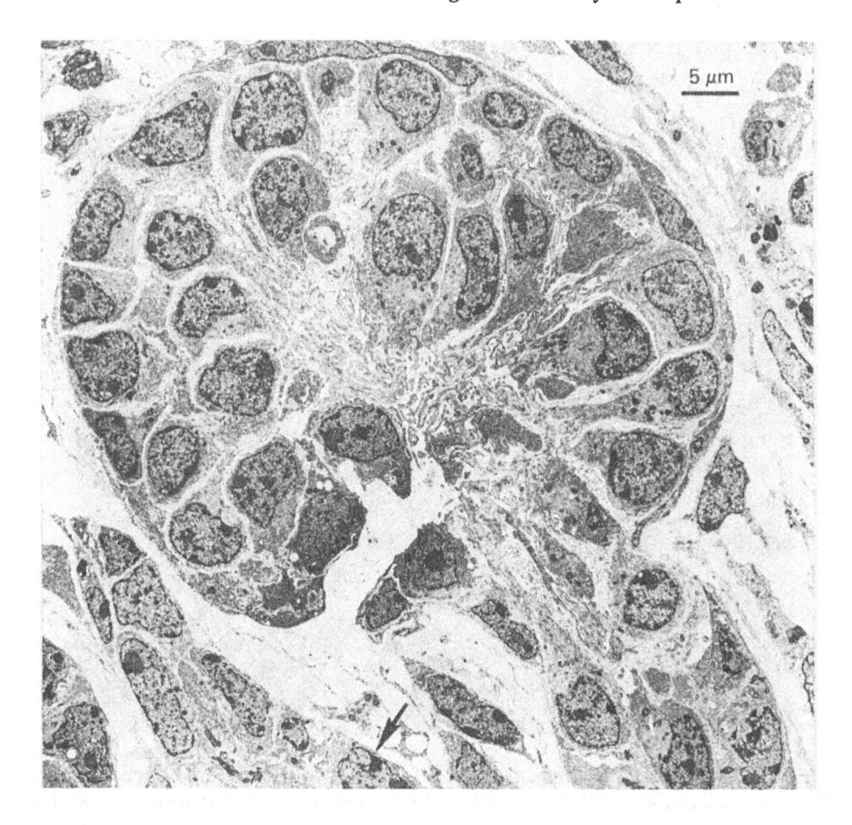

Fig. 4.22. Electron micrograph of a well-shaped glomerular body in a transfilter-induced nephric mesenchyme cultivated for seven days. Attached to the stalk an arborizing basement membrane may be seen. No vascular elements are detected (Bernstein *et al.*, 1981; courtesy of Dr J. Bernstein).

continues over an extended period, but is greatly delayed. The nature of this antigen and its 'indicator significance' were not revealed by this study.

Glomerular differentiation

Direct ultrastructural observations and the appearance of epithelial podocytes which bind colloidal iron onto their surface show that structures with many features common to the primitive glomeruli *in vivo* form in the transfilter-induced cultures (Bernstein *et al.*, 1981). Further evidence has been obtained by the use of fluorochrome-conjugated lectins binding to known sugar moieties on the surface of the podocytes. Wheat germ agglutinin (WGA) binds to sialic acid and peanut agglutinin (PNA) to galactose residues. When sections of whole embryonic kidneys were treated with fluorescein-conjugated WGA, a rather selective bind-

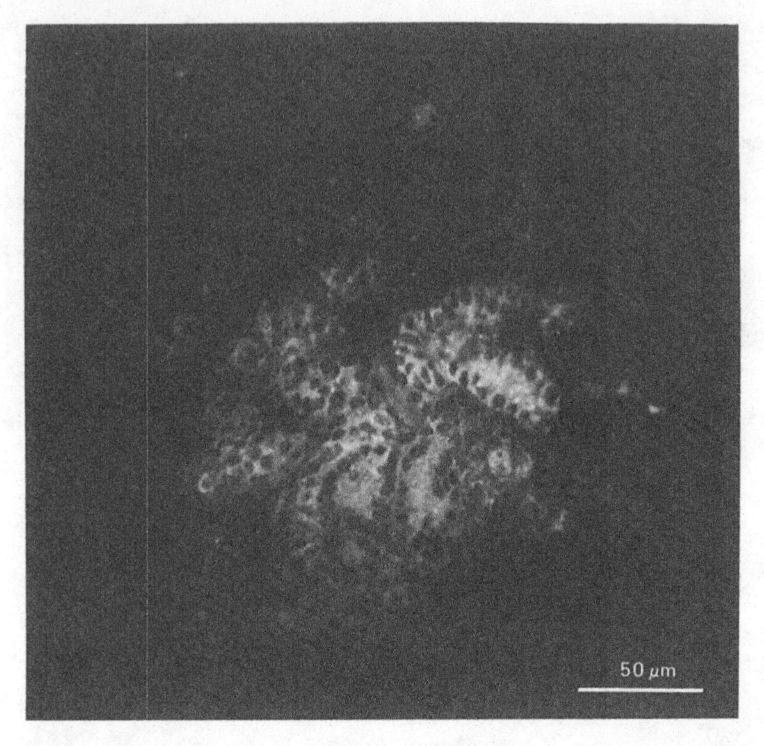

Fig. 4.23. Fluorescence micrograph of tubules in a transfilter-induced mesenchyme cultivated for 12 days and reacted with a kidney-specific antigen (Lahti & Saxén, 1966).

ing to the anionic podocyte surface was obtained on day 13. PNA bound to these surfaces only irregularly or not at all, but after the sialic acid moieties were removed by neuraminidase treatment, the WGA-binding sites disappeared, and the PNA-binding sites were unmasked, resulting in strong and regular binding (Ekblom *et al.*, 1981*a*; Laitinen *et al.*, 1986). This rather early segregation of the glomerular epithelium *in vivo* can be reproduced *in vitro*. In the transfilter-induced mesenchymes, WGA-binding sites could be visualized on the third day of culture, and, correspondingly, PNA binds to the primitive glomerular bodies after neuraminidase treatment (Ekblom *et al.*, 1981*a*). The binding of colloidal iron to these surfaces, as reported by Bernstein *et al.* (1981), has also been confirmed independently (Lehtonen *et al.*, 1983). Indirect evidence for the glomerular nature of the primitive, avascular bodies developing in the experimentally induced mesenchymes *in vitro* is presented in detail in Chapter 5. Briefly, when mesenchymes developing such glomerular bodies are grafted onto chick chorioallantoic membranes, they form rather advanced glomeruli with a vascular component provided by the avian vessels (Sariola *et al.*, 1983).

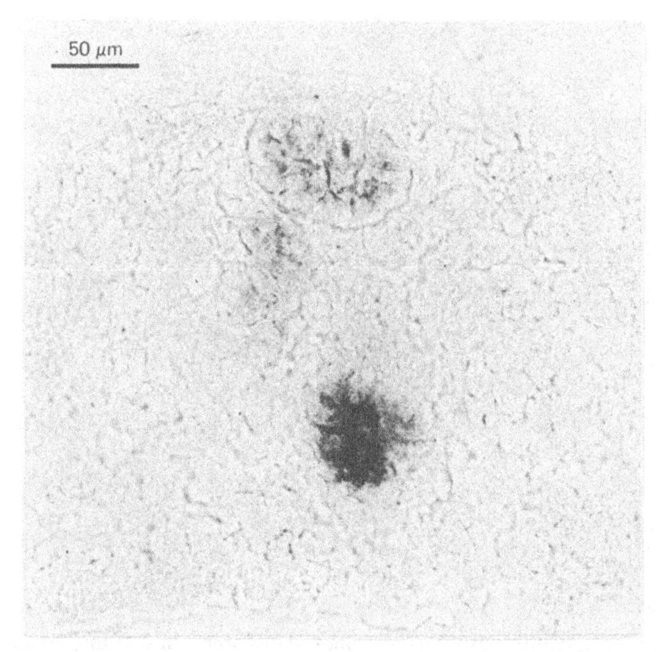

Fig. 4.24. Micrograph of the enzyme γ-glutamyltransferase detecting the proximal tubules in a transfilter-induced mesenchyme (Lehtonen *et al.*, 1983; courtesy of Dr E. Lehtonen).

Proximal tubules

In order to examine the possible segregation *in vitro* into proximal and distal tubules of the nephron, both immunohistological, histochemical and electron-microscopic techniques have been used. An antibody against the brush border (BB) of the rat proximal tubules was prepared by Miettinen & Linder (1976) and shown to cross-react with mouse tissues. In the transfilter cultures, the first binding of the anti-BB-antibody to the apical surfaces of the tubules was obtained on the fourth day of cultivation (Ekblom *et al.*, 1980*b*, 1981*a*). The enzyme γ-glutamyltransferase (GGT) has been considered to be a specific marker for the proximal tubules (Rutenberg *et al.*, 1969), and our observations on sections of whole embryonic kidneys lend further support to this view. In the transfilter cultures GGT-activity could be demonstrated in many tubules on the fourth day of cultivation, concomitant with, or slightly later than, the appearance of the BB antigen (Lehtonen *et al.*, 1983; Fig. 4.24). By electron microscopy, tubules were observed on day 5 characterized by numerous microvilli at the apical cell surfaces. These microvilli probably represent the brush border of the proximal tubules (Ekblom *et al.*, 1981*a*; Lehtonen *et al.*, 1983; Fig. 4.25).

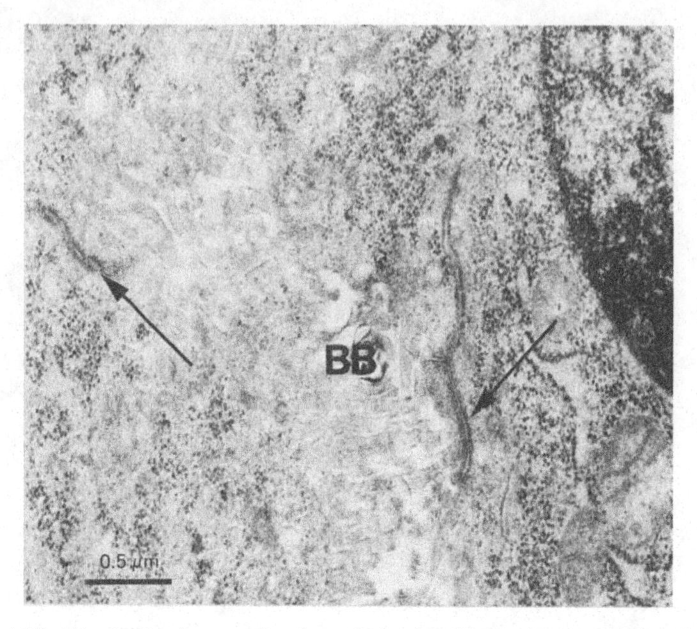

Fig. 4.25. Electron micrograph of a tubule developing in a transfilter-induced mesenchyme. Microvilli suggestive of the brush border of the proximal tubules are visualized (BB). Intercellular junctions are indicated by arrows (Lehtonen *et al.*, 1983; courtesy of Dr E. Lehtonen).

Distal tubules

An antigen prepared against the Tamm–Horsfall (TH) protein was applied to follow the possible appearance of distal tubules in the cultures. This glycoprotein has been found in mammalian embryonic kidneys, and it is considered to be restricted to the distal portion of the nephron (Hoyer *et al.*, 1974; Sikri *et al.*, 1979; Dawnay *et al.*, 1980). *In vivo*, the appearance of this antigen seems to lag behind the first occurrence of the BB antigen, and, correspondingly, in the cultures of induced metanephric mesenchymes, the first TH-positive tubules could be detected by immunohistology on day 5, i.e. some 24 h after the appearance of the BB antigen (Ekblom *et al.*, 1981*a*).

Electron microscopy of mesenchymes on day 5 in culture revealed some tubules devoid of the microvilli shown in Fig. 4.25. These tubules consisted of uniform, polarized cells with a distinct basement membrane but no protrusions at their apical surfaces (Fig. 4.26). They were suggested to be tubules of the distal portion of the nephron (Lehtonen *et al.*, 1983).

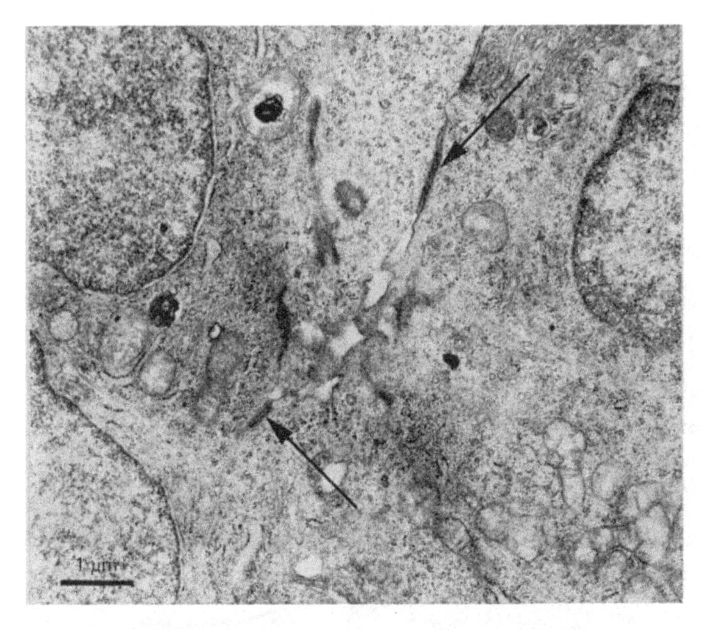

Fig. 4.26. Electron micrograph of a tubule in an induced mesenchyme on day 5. The uniformly shaped, polarized epithelial cells are devoid of microvilli and suggestive of a distal tubule. Intercellular junctions are indicated by arrows (Lehtonen *et al.*, 1983; courtesy of Dr E. Lehtonen).

Induction of segmentation

Observations on the transfilter-induced metanephric mesenchyme show that induction is followed not only by early aggregation and epithelialization of the mesenchymal cells and their assembly into nephric vesicles, but also by a rather advanced differentiation of the cells expressing phenotypes characteristic of the three major segments of the nephron. The results also suggest that the segmental differentiation is sequential and follows the order glomerulus $\rightarrow$ proximal tubules $\rightarrow$ distal tubules (though one should use caution when comparing results obtained with different probes and with techniques that are not necessarily of the same sensitivity). The most intriguing question that then arises is how this differentiation programme is implemented. As shown in the previous chapter (Fig. 3.18, p. 81), the induction of kidney tubules in the experimental set-up is completed after some 24 to 26 h, and then the mesenchymal cells assemble into tubules after a 'morphogenetically silent' period of 12 to 16 h (Saxén & Lehtonen, 1978). To examine whether this relatively short induction 'pulse' would also suffice to programme the cells for further segmentation and expression of segment-specific markers as just described, experiments were conducted with the technique illustrated in Fig. 3.16 (p. 79): the mesenchymes were induced

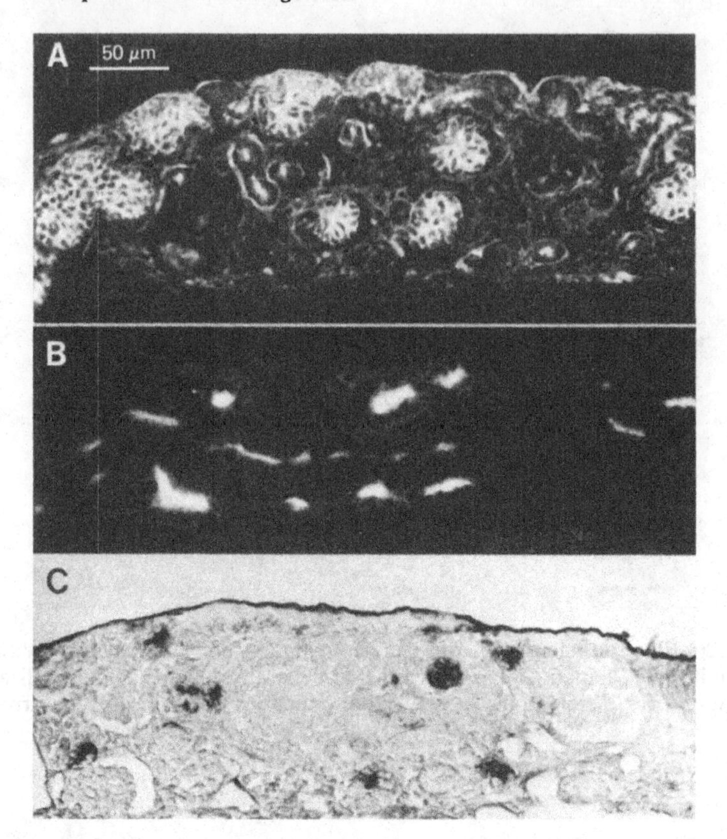

Fig. 4.27. Illustration of the results of an experiment where the metanephric mesenchyme was induced transfilter for 24 h and subcultured for five days (Ekblom *et al.*, 1980*b*, 1981*a*). A. Fluorescence electron micrograph of a section treated with fluorochrome-conjugated wheat germ agglutinin. Advanced glomerular bodies are detectable. B. Fluorescence micrograph of a section reacted with anti-brush-border antibody visualizing the proximal tubules. C. Micrograph of a section treated with anti-Tamm–Horsfall antibody detecting the distal tubules. The peroxidase technique was used.

for 24 h by a transfilter contact with the spinal cord and subcultured for an additional three to five days and then examined with the techniques described earlier in this chapter. It became evident that the short, 24-h induction pulse did programme the entire segmentation of the nephron, and when examined on day 5, all three major segments of the nephron could be visualized with the markers used (Ekblom *et al.*, 1981*a*; Lehtonen *et al.*, 1983; Fig. 4.27). When the induction time was shortened further, the observations became hampered by technical difficulties; only a few tubules developed as expected (Saxén & Lehtonen, 1978), and their viability in the gradually disintegrating, uninduced mesenchyme was low. Thus far, all tubules analysed in such cultures have shown the BB antigen and the somewhat less specific GGT reactivity, but not the TH antigen of the distal tubules. Features typical of the developing glomerular bodies *in vitro* were also regularly detected in such cultures. It is tempting to suggest that a sequential triggering occurs of the three cell lineages that express the characteristics of the different segments of the nephron. Evidence for this postulate, however, is still insufficient (Lehtonen *et al.*, 1983).

Epithelial Phenotype *versus* Tubulogenesis

Introduction

The complex chain of events initiated in the nephric mesenchyme by an interaction with the epithelial inductor involves expression of new cellular phenotypes but also an assembly of these cells into specific, three-dimensional structures. Clearly, events at the cellular and tissue levels are interdependent, and after induction the pretubular cells interact with each other ('homotypic' interaction, Grobstein, 1962) and also with their extracellular matrix (basement membrane). To understand the control mechanisms of this complicated organogenetic process, an effort should be made to separate events at the cellular and tissue levels and to examine to what extent the expression of a new cell phenotype is a consequence of spatial assembly rather than independent cytodifferentiation at the level of single cells. Two previous approaches along these lines in other embryonic organs should be mentioned. Spooner *et al.* (1977) followed the onset and development of a specific enzyme (amylase) activity in pancreatic epithelial cells normally forming exocrine acini. Prevention of this assembly into acini in monolayer cultures did not inhibit cytodifferentiation of the predetermined epithelial cells. This was shown as a 100-fold rise in amylase activity and as appearance of zymogen granules.

Correspondingly, the endocrine activity of Leydig's cells of the testis normally develops within the seminiferous cords. Formation of these cords can be prevented *in vitro* by foetal calf serum, and yet cells releasing testosterone appear within the disorganized testicular rudiments (Magre & Jost, 1984; Patsavoudi *et al.*, 1985).

Tubulogenesis of the kidney has recently been subject to a corresponding analysis by Lehtonen and co-workers (Lehtonen *et al.*, 1985, and unpublished results; Lehtonen & Saxén, 1986*b*).

Monolayer cultures

To dissect cytodifferentiation (epithelialization) and morphogenesis (tubulogenesis) in the embryonic kidney, the transfilter model-system was combined with Spooner's strategy (Spooner *et al.*, 1977). With the knowledge that, in a transfilter culture, at 24 h a great proportion of the mesenchymal cells are determined towards an epithelial destiny, such mesenchymes were examined in monolayer cultures. After induction, the still fully undifferentiated mesenchymes were either chopped into tiny pieces or gently dissociated by enzyme treatment – in both cases monolayer cultures developed on coverslip surfaces. Cells in such culture conditions showed no spatial arrangement, they divided mitotically and soon showed some morphological diversification.

To follow developmental changes in the monolayer cultures, immunofluorescence techniques with specific antibody probes were used. Two of the antibodies were against cytoskeletal proteins (vimentin, cytokeratin) and two against compounds of the extracellular matrix (fibronectin, laminin). I have already described the changes in the expression of these proteins *in vivo* and in the three-dimensional model-system: uninduced mesenchymal cells of the blastema express vimentin, a common marker for embryonic mesenchymal cells, but lose these filaments when converted into cytokeratin-positive epithelial cells. Similarly, fibronectin is expressed by undifferentiated mesenchymal cells but disappears from the matrix after induction, remaining in the inter-tubular stroma. Conversely, laminin synthesis is enhanced in the induced mesenchymal areas, after which it is confined to the basement membrane of the early renal vesicles. Dispersed cells in monolayer cultures followed this developmental pattern only partially.

Results obtained in the monolayer cultures can be summarized as follows (Lehtonen & Saxén, 1986*b*; Figs. 4.28 to 4.30).

(1) Conversion of cells expressing vimentin-type intermediate filaments into epithelial cells with cytokeratin-positive filaments occurred regularly in many but not all cells in the induced cultures. 'Intermediate' cells with both types of filaments were not rare.

(2) Expression of the two proteins of the extracellular matrix did not, in contrast, follow the developmental pattern *in vivo*; synthesis and deposition of both laminin and fibronectin were observed throughout the monolayer and in cultures of uninduced cells. No clear differences, spatial organization or polarized secretion of these proteins were recorded. Furthermore the expected co-distribution vimentin–fibronectin and cytokeratin–laminin was not distinct, and both vimentin- and cytokeratin-positive cells synthesized the two matrix proteins (Figs. 4.29 and 4.30).

The above results and their comparison with our earlier data on development *in vivo* and changes in the three-dimensional model-system have led us recently to the following tentative conclusions (Lehtonen & Saxén, 1986b):

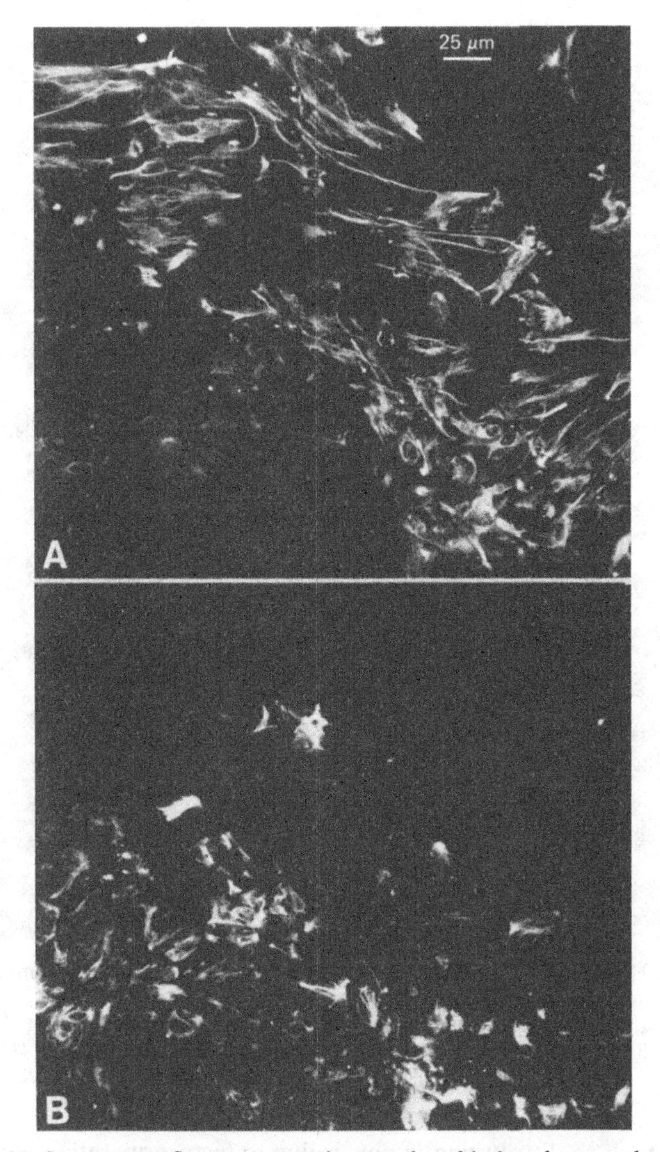

Fig. 4.28. Low-power fluorescence micrographs of induced mesenchymal cells cultivated as monolayers and double stained with antisera against vimentin (A) and against cytokeratin (B) (Lehtonen *et al.*, 1985).

(1) Induction of the nephric mesenchymal cells results in an assembly-independent change in phenotype (as shown by the two markers for the intermediate filaments).
(2) Synthesis of laminin and fibronectin is inherent in uninduced cells,

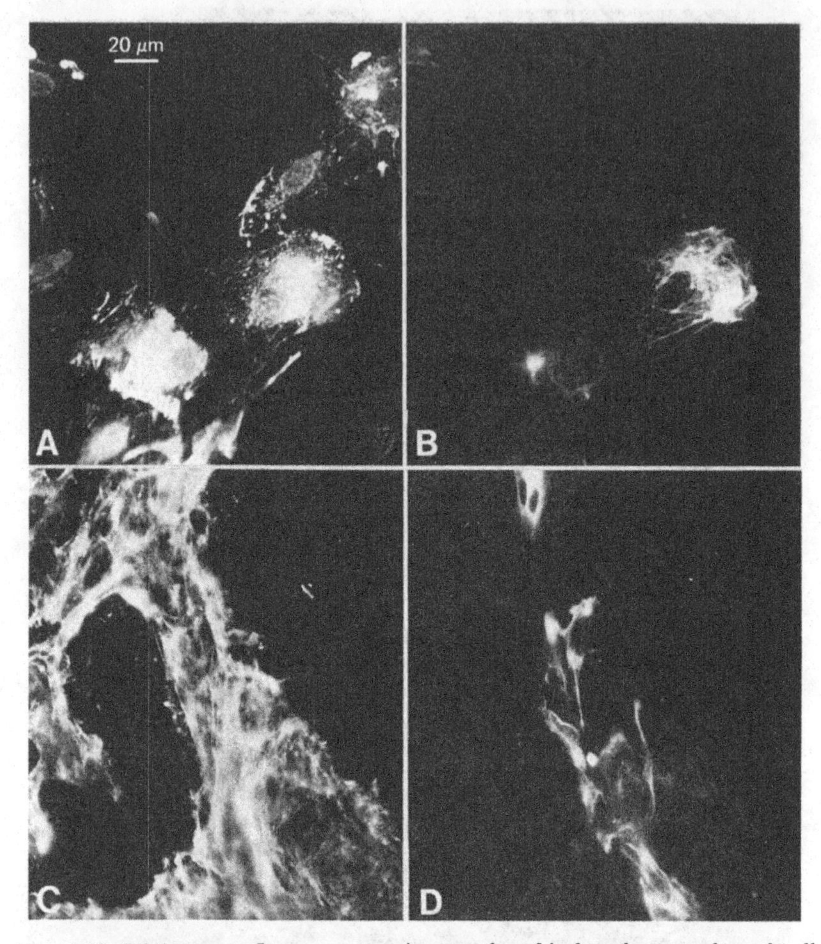

Fig. 4.29. High-power fluorescence micrographs of induced mesenchymal cells cultivated for seven days as monolayers and double stained with two antisera. Staining of fibronectin (A) and cytokeratin (B) reveal a partial co-distribution with a cell expressing both antigens. Double staining for laminin (C) and cytokeratin (D) show several cells expressing both proteins (Lehtonen & Saxén, 1986*b*)

and their re-distribution *in vivo* is due to factors operating at tissue level, e.g. the changing architecture, local activation of proteolytic enzymes and organismal control mechanisms.

(3) Polarization of cells is related to their three-dimensional assembly and interaction with extracellular components.

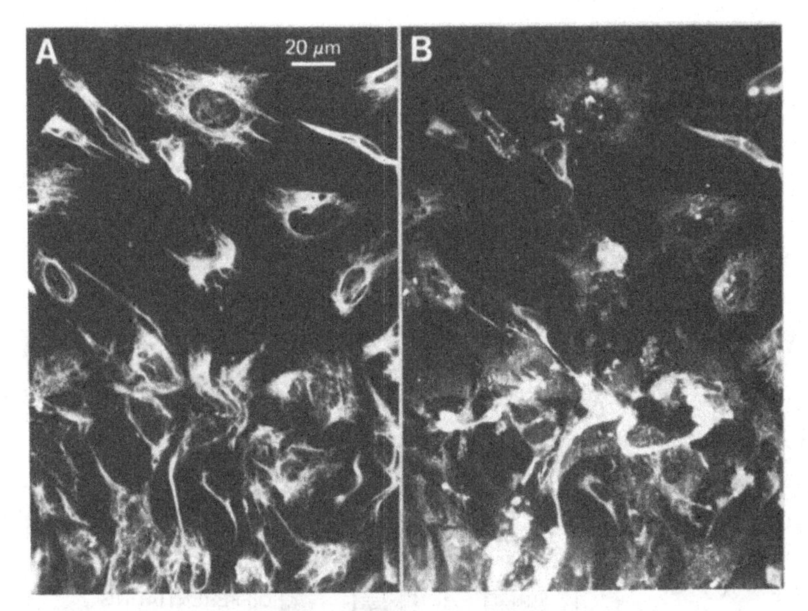

Fig. 4.30. High-power fluorescence micrographs of mesenchymal cells induced for 24 h and subsequently cultivated as monolayers for five days. Double staining for laminin (A) and vimentin (B) show partial co-distribution (Lehtonen & Saxén, 1986*b*).

Summary of kidney tubule differentiation *in vitro*

Our present knowledge of the induction of kidney tubules and of the early stages of tubule formation *in vitro* are summarized in Fig. 4.31. Because a majority of the observations are derived from experiments that have been made using the same basic techniques and experimental conditions, a temporal correlation of the various events is possible. It is likely though that the scheme will be modified and completed in the future, when new and more sensitive techniques become available for detecting early changes at the submicroscopic and molecular levels. The course of the differentiative changes may also allow some conclusions and stimulate speculations on their causal relationship. This will be done in Chapter 6.

The induction of the metanephric kidney tubules is permissive and the inductor acts upon a predetermined cell population with restricted developmental options. Induction of the mesenchymal cells is a rather slow event and is completed only some 24 h after an inductor/target tissue contact has been established. This induction 'pulse' leads to an epithelial transformation of the mesenchymal cells and to their assembly into tubular structures. Induction, furthermore, programmes the target cells ultimately leading to the expression of several cellular phenotypes within the nephron. The inductive signals and the mechanisms of the programming are unknown.

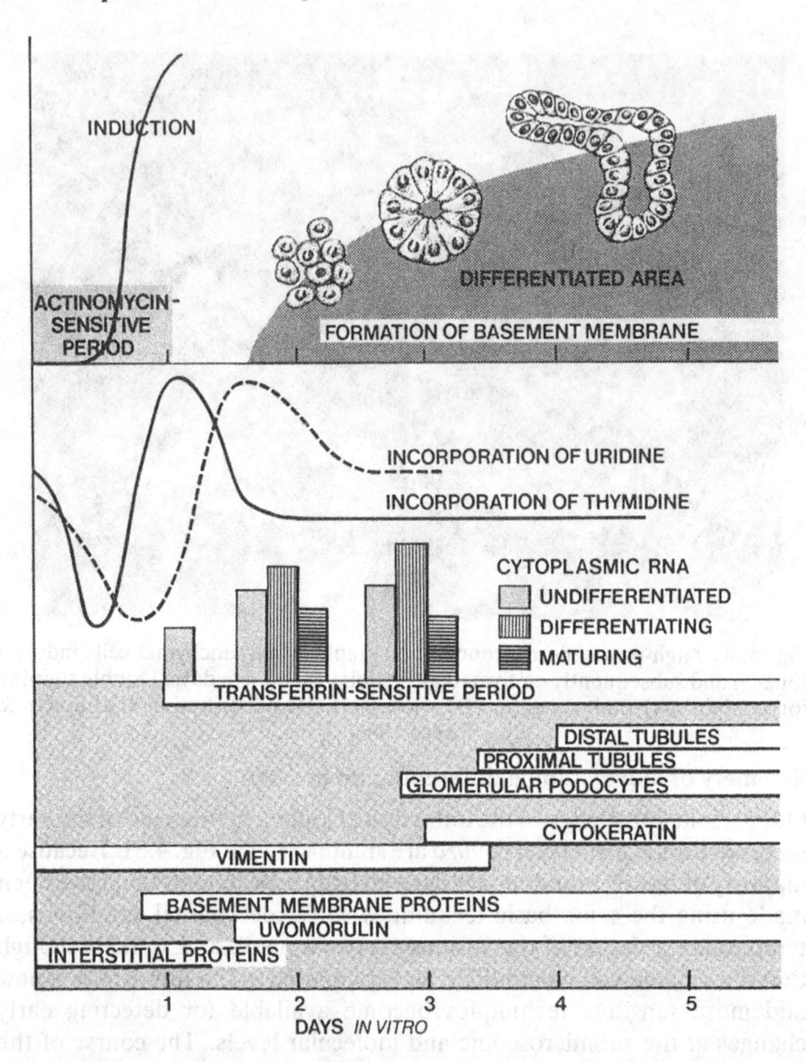

Fig. 4.31. Scheme of the various parameters monitoring the differentiation of the metanephric mesenchyme in transfilter culture. Modified from diagrams published by Saxén *et al.* (1968, 1981), Saxén (1971), Ekblom (1984) and Lehtonen & Saxén (1986*a*).

The various changes and differentiative parameters summarized in Fig. 4.31 may justify the distinguishing of five periods of kidney tubule induction and differentiation in the experimental model-system. No doubt, the division is schematic and the periods partially overlapping:

Period 1 begins with the establishment of the inductor/mesenchyme contact, which takes place 1 to 2 h after the setting up of the transfilter culture. The period lasts for some 12 h and is characterized by a low metabolic activity of the cells without any detectable changes connected with differentiation. The gradually decreasing incorporation rate of precursors for macromolecules apparently reflects a slow recovery from the preceding mechanical manipulation and enzymic treatment of the mesenchymal cells.

Period 2 covers hours 12 to 24, when rapid changes in the mesenchyme can be monitored by several parameters. Synthesis of DNA and RNA shows an increasing trend as recorded by the incorporation of their labelled precursors and by the measurement of the cytoplasmic RNA. Parallel to this increase, which reaches its climax at 24 h, the mesenchyme loses its sensitivity to certain metabolic inhibitors such as actinomycin D. The induced cells become sensitive to the growth factor transferrin which from this point is a necessary constituent of the culture medium.

The extracellular matrix undergoes marked changes during the second period of induction. Initially the mesenchymal extracellular matrix contains interstitial proteins (collagen type I and type III, fibronectin), but towards the end of the second period these disappear, as judged by immunohistology. A set of 'epithelial' proteins can then be visualized with the same technique, and this apparently reflects their enhanced synthesis. These proteins include collagen type IV, laminin, and a heparan-sulphate-rich proteoglycan. No definite morphological changes occur during this period, but time-lapse cinematography suggests that there is already a decreased motility of the mesenchymal cells at this stage.

Period 3 covers the following 12-h period from 24 to 36 h when induction is actually completed. Synthesis of DNA slowly diminishes, but remains clearly above the level of uninduced controls. Towards the end of this period the epithelializing cells can be distinguished from the uninduced cells by their aggregation into pretubular condensates. Morphologically, the first signs of an accumulation of basement-membrane-like material are seen around the condensates, and immunohistology shows how the epithelial proteins become confined to the periphery of the condensates. Cells soon become attached to this basement membrane, which initiates the fourth period.

Period 4 begins with the attachment of cells to the newly formed basement membrane, followed by their polarization. A central lumen opens, and electron microscopy reveals a definitely polarized ultrastructure. Cells begin to lose their vimentin-type intermediate filaments and start expressing cytokeratin.

Period 5 is characterized by the elongation and irregular coiling of the nephric tubules – this would correspond to the formation of the **S**-shaped body *in vivo*. The total induced area of the mesenchyme is still increasing. Towards the end of this period, which lasts for three to five days, segregation of the nephron can be shown and three new epithelial cell phenotypes are expressed in a sequential manner: the presumptive glomerular podocytes binding certain lectins and colloidal iron on their anionic surface on day 3; the brush border of the proximal tubules can be visualized by immunohistology and by electron microscopy on day 4; and, finally, the TH protein restricted to the distal tubules is expressed on day 5. Needless to say, the period of maturation extends beyond these stages as new cell types differentiate *in vivo*. However, *in vitro*, the sequence might end with these first signs of segmentation of the nephron.

5
Vascularization of the nephron

Introduction

The origin of the endothelial component of the glomerulus and the mode of development of the kidney vasculature have been controversial subjects for a century. The original hypotheses were outlined by embryologists who based their ideas on light-microscopic findings. Later, electron-microscopic explorations could not settle the varying opinions. The somewhat confusing literature has been reviewed by Jokelainen (1963), Potter (1965), Kazimierczak (1971) and Sariola (1985).

Three cell lineages are theoretically potential progenitors for the capillary endothelial cells of the glomeruli, and all these possibilities have found their adherents. Accordingly, the following possible origins can be listed:

(1) Cells of the glomerular anlage, the renal epithelium, derived from the nephrogenic mesenchyme.
(2) Mesenchymal cells, either within the glomerular crevice or invaginating there.
(3) Endothelial cells outside the nephric blastema.

The first hypothesis, the origin *in situ* of the endothelial cells, was first suggested by Herring (1900). Similar views were presented by Rienhoff (1922) and, especially, by Hall & Roth (1957). The two latter authors observed red blood cells within the glomerular body before capillary lumina were detected. They considered this to be further evidence for the endothelial and hematopoietic potentials of the glomerular cells before the development of outside vessels and connections to them. The finding was confirmed by Kurtz (1958), who also agreed with the conclusions drawn by Hall & Roth of the origin *in situ* of the hematopoietic and the endothelial cells. This view was shared by Lewis (1958) and Suzuki (1959) and in 1962 by Vernier & Birch-Anderson, who concluded that 'buds of renal artery join the (glomerular) tuft only after formation of a vascular network *in situ* within the glomerulus'.

In his classic paper in 1905, Huber suggested that the glomerular vasculature might be derived from cells outside the S-shaped anlage: 'My own observations lead me to think that the first capillary loop found

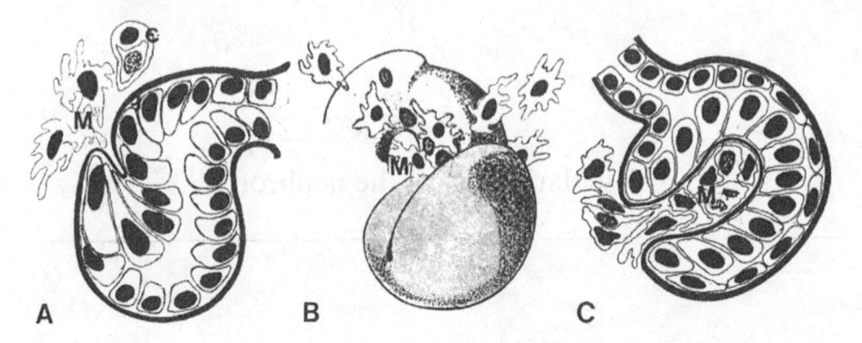

Fig. 5.1. Scheme of vasculogenesis of the nephric vesicle as seen in serial thin sections of immature rat kidney (Jokelainen, 1963). A. Formation of the lower (glomerular) crevice at the comma-shaped stage. Mesenchymal cells (M) and stromal capillaries (c) are seen around the orifice. B. Reconstruction of the S-shaped body with mesenchymal cells invading the crevice and a row of erythrocytes. C. Section through the lower crevice of an immature glomerulus showing the invaded cells within the glomerular crevice.

within the mesenchyme occupying the concavity of the lower S-curve of a tubular anlage grows into this from without as I have generally been unable to trace a connection between it and the capillaries outside the cleft occupied by the mesenchyme.' Though briefly mentioned by Edwards (1951) and Suzuki (1959), the 'migration *in situ*' hypothesis was forgotten until Jokelainen (1963) presented further light- and electron-microscopic evidence for it. In his serial thin sections he found mesenchymal cells which 'migrate actively' into the glomerular cleft and become differentiated into endothelial cells (Fig. 5.1). He also described primitive capillaries around the orifice of the cleft, which frequently contained red blood cells. Jokelainen emphasized that these erythrocytes were regularly mature and that no hematopoiesis was detected. While adhering to the 'migration *in situ*' hypothesis, he did not, however, explain where these apparently circulating cells came from. In his thorough ultrastructural analysis of early glomerulogenesis, Kazimierczak (1965, 1970, 1971) emphasized that the morphology of the endothelial cells showed that they were motile and invaded the glomerular crevice. He also stressed that, through glomerulogenesis, the capillaries within the S-shaped body were connected to the 'renal vascular blood system', but he did not define clearly how the latter was formed. This idea of the origin of the glomerular epithelium was shared by Aoki (1966).

The third hypothesis suggesting an outside origin for the kidney vasculature was first stated by Potter in 1965. Basing her observations on direct light microscopy and lengthy microdissection she concluded: 'The capillaries originate as direct outgrowths from adjacent vessels and the cells they contain come from the general circulation . . . and there is no evidence that capillaries exist in a glomerulus at any stage of development that are not in direct communication with the vessels outside the glomerulus' (Potter, 1965; Osathanondh & Potter, 1966). Despite these statements and direct and circumstantial evidence (e.g. the mature red cells within the glomeruli), the '*in situ*' hypothesis has persisted (Reeves *et al.*, 1980), and even some additional experimental evidence has been provided (below).

Experimental investigations

Direct evidence for a vasculopoietic and hematopoietic competence of the metanephric mesenchymal cells (or of the induced epithelial cells) would be their conversion *in vitro* into endothelial or blood cells. In fact, this has been reported by Emura & Tanaka (1972). Using the transfilter technique, described on p. 48, they combined the undifferentiated, murine nephric mesenchyme with embryonic liver and obtained both endothelial lacunae and blood cells in the former. The obvious conclusion was that the mesenchymal blastema contained cells with vasculogenetic and hematopoietic developmental potentials. We have not, however, been able to confirm these important observations. Using an almost identical technique and the same tissue components we have not observed endothelial or hematopoietic cells within the metanephric mesenchyme (Sariola *et al.*, 1982). An epithelial origin of the endothelial/hematopoietic cells was excluded in similar experiments where the mesenchyme was combined with both the liver and an inductor. This led, as expected, to the formation of epithelial tubules but never any other cell types.

Several explanations can be offered for the discrepancy between our results and those of Emura & Tanaka (1972). Our culture conditions, differing slightly from those of the Japanese authors, might not support endothelial/hematopoietic differentiation, or Emura & Tanaka might have used mesenchymes contaminated by other cells. There may be two sources of such cells: at the time of dissection the metanephric blastema is surrounded by a capillary network which could provide endothelial cells to the explant, especially if slightly 'aged' rudiments are used. Another source of contaminating cells could be the liver tissue at the lower surface of the filter. The irregular, often rather large pores of filters that were in use in 1972 could have allowed passage of hematopoietic/endothelial cells from the liver to the mesenchymal component.

Indirect experimental evidence for the non-nephrogenic origin of the glomerular endothelium is gained from long-term organotypic cultures of embryonic kidney rudiments and transfilter-induced mesenchymes. Both develop rather advanced epithelial structures including glomerular

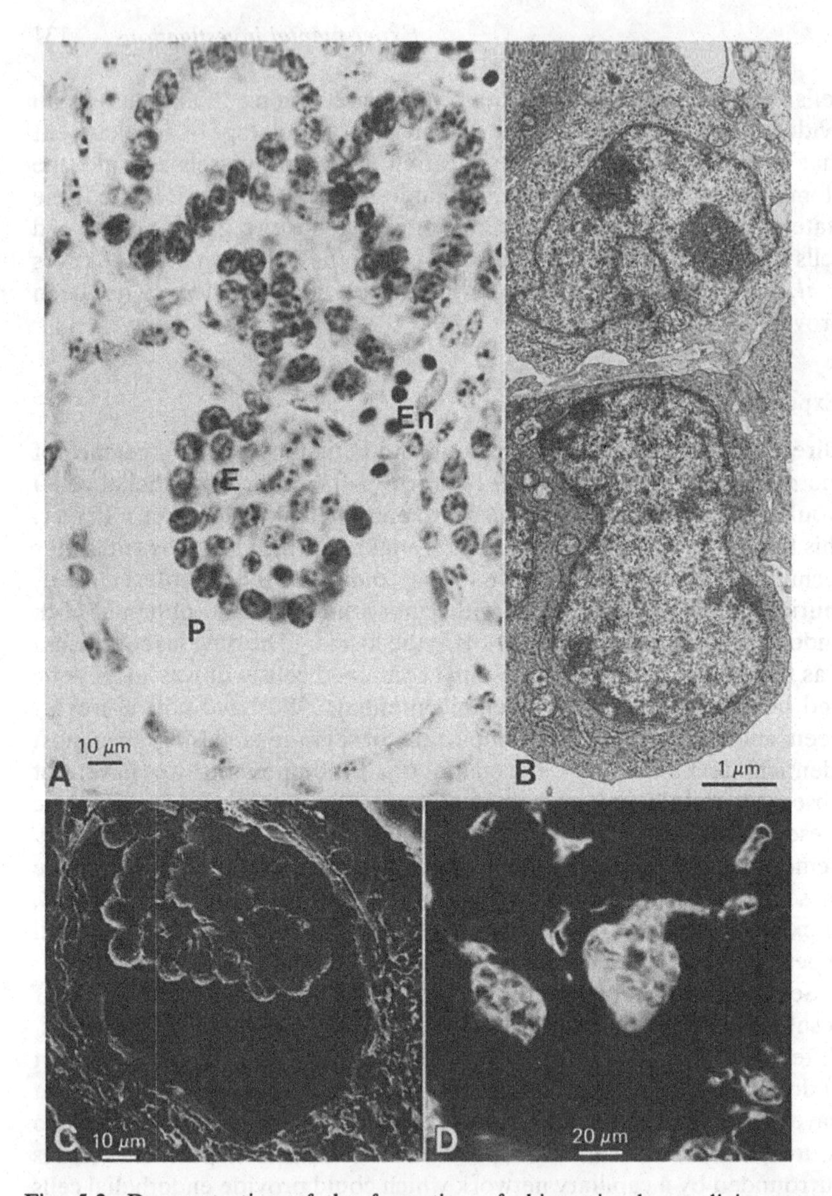

Fig. 5.2. Demonstration of the formation of chimaeric glomeruli in mouse kidneys grafted onto quail chorioallantoic membrane (Sariola *et al.*, 1983; Sariola, 1985; courtesy of Dr H. Sariola). A. Micrograph of a Feulgen-stained section shows quail endothelial cells with deeply stained nucleoli (En) within the glomeruli and erythrocytes (E) at its orifice. The endothelial cells are surrounded by mouse-derived podocytes (P). B. Electron micrograph shows a quail endothelial cell with its nuclear 'marker' and a mouse podocyte. C. Scanning electron micrograph illustrates the folding surface of the glomerular loop. D. Fluorescence micrograph after staining with a monoclonal antibody against quail endothelial and hematopoietic cells also demonstrates the origin of the glomerular endothelium.

bodies, but these remain avascular, as reported by two independent laboratories (Bernstein, 1978; Bernstein *et al.*, 1981; Ekblom, 1981*b*; Bonadio *et al.*, 1984).

The findings and conclusions concerning the origin of the kidney vasculature are still variable. Some recent experimental studies related to the problem will be described.

Grafting experiments

Previous findings demonstrate that many embryonic and neoplastic tissues elicit a vascular response when grafted onto avian chorioallantoic membrane (CAM) (p. 38) (for a review, see Folkman & Cotran, 1976). These tissues include the embryonic kidney, as might have been expected from previous transplantation experiments demonstrating that embryonic and neonatal murine kidneys give a vascular response in their new sites (Grobstein & Parker, 1958; Warren *et al.*, 1972). Auerbach *et al.* (1976) grafted undifferentiated mouse kidney rudiments onto CAM and obtained no neovascularization, but a strong vascular response was detected with older kidney tubules.

We have confirmed the above finding by Auerbach *et al.* (1976). Undifferentiated metanephric mesenchymes remained avascular on CAM, whereas differentiating kidney rudiments and transfilter-induced mesenchymes became richly vascularized (Ekblom *et al.*, 1982; Sariola *et al.*, 1983; Fig. 5.2). Light- and electron-microscopic examination of the glomerular anlagen that develop in these conditions reveal an endothelial component, rare in the induced mesenchymes but regular in whole-kidney grafts.

For further analysis of the origin of the glomerular endothelium, we used the quail 'marker' described by Le Douarin (1973; Le Douarin & Barq, 1969) and illustrated in Fig. 3.26 (p. 87). Mouse kidneys were grafted onto quail CAM, and the glomerulus-bearing grafts were analysed either by Feulgen-staining in sections or by electron microscopy. Both techniques revealed conclusively that the glomerular endothelium and the kidney vessels were of quail origin (Ekblom *et al.*, 1982; Sariola *et al.*, 1983; Fig. 5.2A,B). 'Contaminating' mouse endothelial cells were found in fewer than 5% of the grafts. In addition, adult quail-type red cells were seen regularly in the vessels and in the glomerular loops. The origin of these as well as of the endothelial cells could be confirmed by using antibodies specific for the quail endothelial and hematopoietic cells (Fig. 5.2D).

Formation of chimaeric murine/avian glomeruli was obtained when whole embryonic kidneys were used as grafts. The results were different when transfilter-induced mesenchyme pieces were similarly grafted and examined: following an initial induction pulse of 24 h or less, areas of the pieces remained uninduced and undifferentiated without any epithelial tubules. In these cases the invading avian vessels seemed to avoid the undifferentiated areas while seeking their way to the epithelialized portions of the explant, where they surrounded the tubular anlagen. No vascularization was observed in uninduced control grafts. Another difference between the whole kidney graft and the transfilter-induced mesenchymes has already been mentioned: in whole-kidney grafts vascularized, chimaeric glomeruli were abundant, whereas such formations were rare in the isolated mesenchymes dominated by avascular glomerular bodies.

Some conclusions can be drawn from the above experimental results. Vessels outside the avascular kidney rudiments are stimulated to sprout and to invade the kidney, where their endothelial cells ultimately home into the epithelial glomerular anlagen to form chimaeric murine/avian glomeruli. The stimulation and attraction are associated with the epithelial differentiation of the kidney mesenchyme, as uninduced mesenchymal areas remain avascular. It is tempting to speculate that induction and subsequent differentiation of the mesenchyme produce an attractant for angiogenesis, an angiogenic factor described in many neoplastic and normal tissues (for a review, see Gullino, 1981). Interestingly enough, some split products of collagens and fibronectin are among such reported angiogenetic factors, and the release of these products is a likely consequence of the early changes in the extracellular matrix (ECM) of the kidney (see p. 22). Recently, Risau & Ekblom (1986) have isolated and characterized an angiogenetic factor from embryonic mouse kidneys. The factor is differentiation dependent and apparently not synthesized by the uninduced metanephric mesenchyme.

Another feature of the results that deserves a short comment is the rare vascularization of the glomeruli formed in the induced mesenchymes devoid of the ureteric component. Two factors might be involved here: either the renal vesicles and the primitive podocytic glomeruli develop temporally mismatched to the endothelial component, or the irregularly shaped epithelial structures in the artificially induced mesenchymes do not provide a proper scaffold for the invading endothelial cells. In brief, some timing and guiding forces that apparently operate during normal development are eliminated in these grafts of mesenchymes devoid of the branching ureter.

To follow the migration of the endothelial cells, additional experiments were performed on whole-kidney grafts (Sariola *et al.*, 1984*c*). These immunohistological investigations bring us back to the earlier reported postinduction changes in the ECM (p. 22).

Angiogenesis and the extracellular matrix

As recently stated by Sariola (1985), the ECM might be involved in vasculogenesis through different mechanisms. In addition to a direct stimulating effect, the constitution of the ECM and its various components could regulate vessel orientation, differentiation and proliferation of the endothelial cells, and finally, render the cells sensitive to circulating angiogenetic factors (Steward & Wiley, 1981; Feinberg & Beebe, 1983; Schor & Schor, 1983). With this background, an attempt was made to correlate the migration of the avian capillaries to the local changes in the ECM reported in Chapter 1 (Sariola *et al.*, 1984*b*). The invading capillaries were visualized by immunofluorescence with two monoclonal antibodies against quail endothelial (and hematopoietic) cells). The constitution of the ECM was examined by antibodies against fibronectin and laminin, the former being a typical constituent of the uninduced stroma, the latter marking the induced areas and, especially, the epithelial basement membranes. As shown in Fig. 5.3, the invading capillaries show a clear preference for the uninduced areas expressing fibronectin, but omit the induced early condensates expressing laminin. The vessels never invade the epithelial tubules surrounded by a laminin-containing basement membrane at a later stage. In fact, the vessels follow the border between the condensates (later renal vesicles) and the uninduced stroma expressing fibronectin. So, the differentiating nephric tissue apparently provides guidance cues for the orderly migration of the endothelial cells, whereas the interactive mechanisms remain speculative. A possible explanation would be that fibronectin is the decisive molecule as has been suggested by Thiery *et al.* (1982*a*) for another migratory event. Since fibronectin is known to convey adhesive properties to cells and adhesive sites for various components of the basement membrane (surrounding the invading vessels), this suggestion seems feasible but remains unproved. Another possibility would be that the randomly migrating endothelial cells seek areas of least mechanical resistance, thus avoiding the areas of early condensation and epithelial vesicles. Whatever the basic mechanism, the migratory paths of the endothelial cells are created by the invading and orderly branching ureter inducing the changes at the molecular and cellular levels.

Glomerulogenesis

Since the glomerular endothelium is of external origin, the sources of two other glomerular components, the basement membrane and the mesangial cells, should be considered. Recent findings of various investigators warrant some conclusions. The glomerular basement membrane (GBM), located between the polarized podocytes and the endothelial

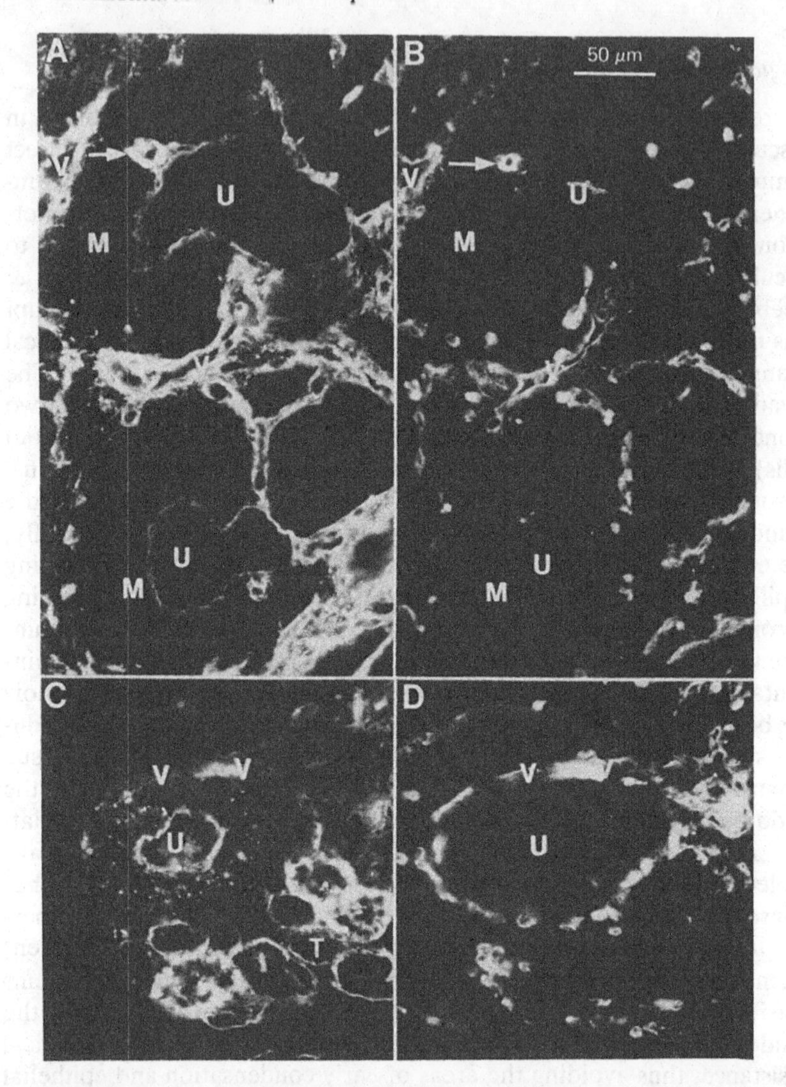

Fig. 5.3. Fluorescence micrographs for comparison of the spatial distribution of fibronectin and laminin with the localization of avian vessels in grafted murine kidneys. Double treatment with antibodies was used against matrix components and quail-derived capillaries (Sariola *et al.*, 1984*c*). A, B. Fibronectin is expressed by the stromal cells, but not by the epithelial cells of the ureteric bud (U) or by the condensed mesenchymal cells (M), and it shows a definite co-distribution with the endothelial cells. C, D. Laminin is expressed in a punctate fashion in the condensates around the ureteric bud (U) and in the basement membrane of the tubules (T), but it shows no co-distribution with the vessels (V).

cells, consists of compounds found in all epithelial basement membranes: collagen type IV and type V, laminin, fibronectin, entactin and heparan sulphate proteoglycan (Ekblom, 1981*a*; Farquhar, 1981; Timpl & Martin; 1982, Avner *et al.*, 1983; Bonadio *et al.*, 1984). Collagen can be localized to the central lamina densa where it forms an elastic cross-linked network, while laminin and heparan sulphate proteoglycan are restricted to the lamina rara, facing both the epithelial (podocytic) and the endothelial (vascular) surfaces (Farquhar, 1981; Timpl *et al.*, 1981; Courtoy *et al.*, 1982). A reasonably good agreement exists between the basic structure and the composition of the GBM, whereas opinions still differ as to its origin.

Since both podocytes and endothelial cells synthesize components of the basement membrane (Jaffe *et al.*, 1976; Killen & Striker, 1979; Gospodarowicz *et al.*, 1981), a dual origin for the GBM has naturally been suggested (Thorning & Vracko, 1977; Huang, 1979; Reeves *et al.*, 1980). Some of the evidence has been derived from transfilter studies *in vitro*, where differentiated avascular glomerular bodies can be explored. The podocytes can be visualized through their polyanionic coat, which binds both (fluorescein-coupled) wheat germ agglutinin and colloidal iron (Bernstein *et al.*, 1981; Ekblom *et al.*, 1981*a*; Lehtonen *et al.*, 1983; Fig. 5.4). After a five-day cultivation of induced mesenchymes *in vitro*, such glomeruli could be visualized, but Ekblom (1981*a*) could not detect any basement-membrane components on the podocytic surface when it was examined by immunohistology. He therefore concluded that endothelial cells might be required for the formation of the glomerular basement membrane. The findings were subsequently revised by another group (Bonadio *et al.*, 1984). By extending the transfilter culture period to seven days, these investigators showed, in controlled immunofluorescence examinations, the expression of collagen type IV and type V, laminin and heparan sulphate proteoglycan at the surface of the visceral epithelial cells (the podocytes) (Fig. 5.5). They concluded that GBM is at least in part an epithelial cell product. Electron microscopy revealed, however, that the basement membrane on the podocytic surface was bilaminar and showed an electron-dense and an electron-lucent layer, but it was devoid of a region corresponding to the lamina rara interna of normal GBM.

The dual original of the glomerular basement membrane has been shown in certain experimental conditions by applying the above-mentioned chorioallantoic grafting technique. The origin of the basement membrane formed in the murine/avian glomeruli was examined with species-specific antibodies against various components of the GBM. The results showed that, in these conditions, the basement membrane contained both mouse- and chick-derived collagen type IV, whereas the basement membrane surrounding the kidney tubules was exclusively of

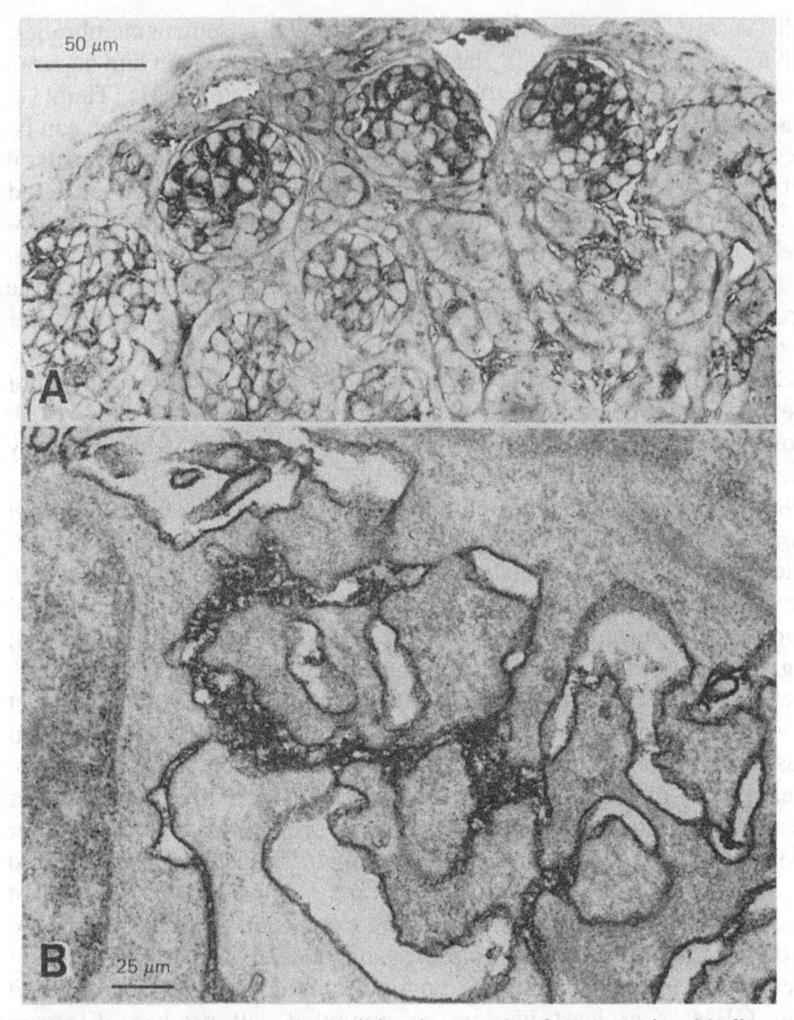

Fig. 5.4. Light (A) and electron (B) micrographs demonstrating binding of colloidal iron on the podocyte surface of glomerular bodies on transfilter-induced mesenchymes (A: Lehtonen *et al.*, 1983; B: Bernstein *et al.*, 1981; courtesy of Dr J. Bernstein).

mouse origin and the one around the vessels of chick type (Fig. 5.6). The authors concluded that the GBM of the chimaeric glomeruli had a dual origin, i.e. it was synthesized by both podocytes and vascular endothelial cells (Sariola *et al.*, 1984*b*). Electron microscopy of these glomeruli showed that the GBM was partially of normal trilaminar type, but frequently split into an epithelial and an endothelial layer, which both showed two regions only, a lamina rara and a lamina densa (Sariola, 1984). Thus, they resembled the GBM of the transfilter-induced

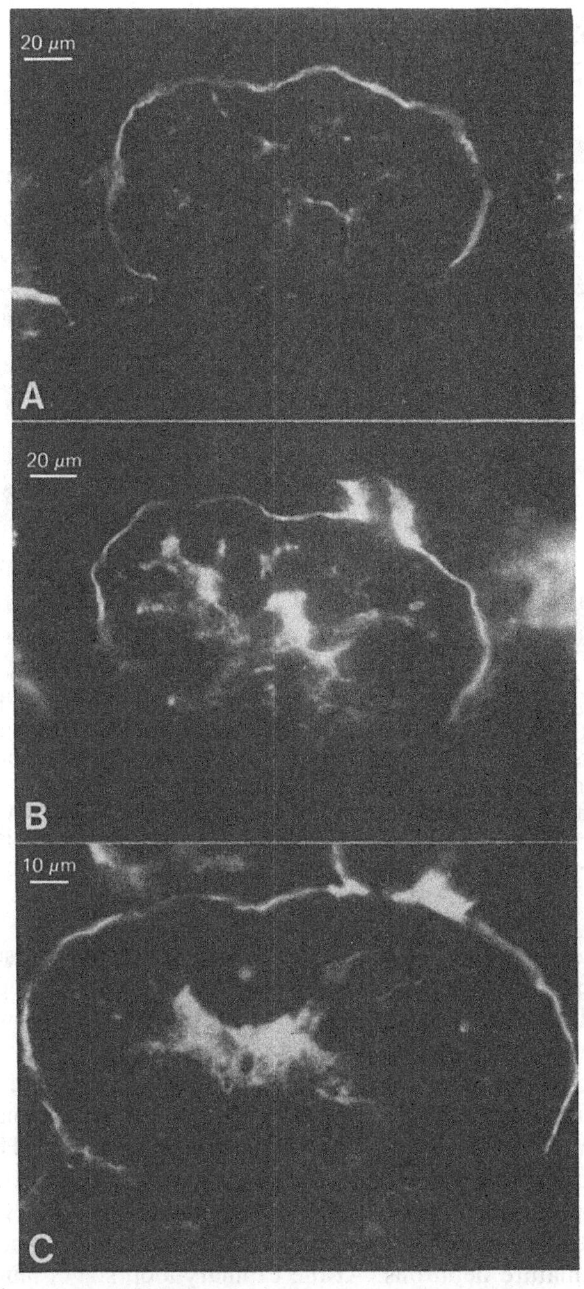

Fig. 5.5. Fluorescence micrographs demonstrating the synthesis of certain basement-membrane components by the podocytes in transfilter-induced mesenchymal explants after seven days of cultivation (Bonadio *et al.*, 1984; courtesy of Dr J. Bernstein). Immunofluorescence technique after treatment with the following antisera: A. Heparan-sulphate proteoglycan. B. Laminin. C. Collagen type V.

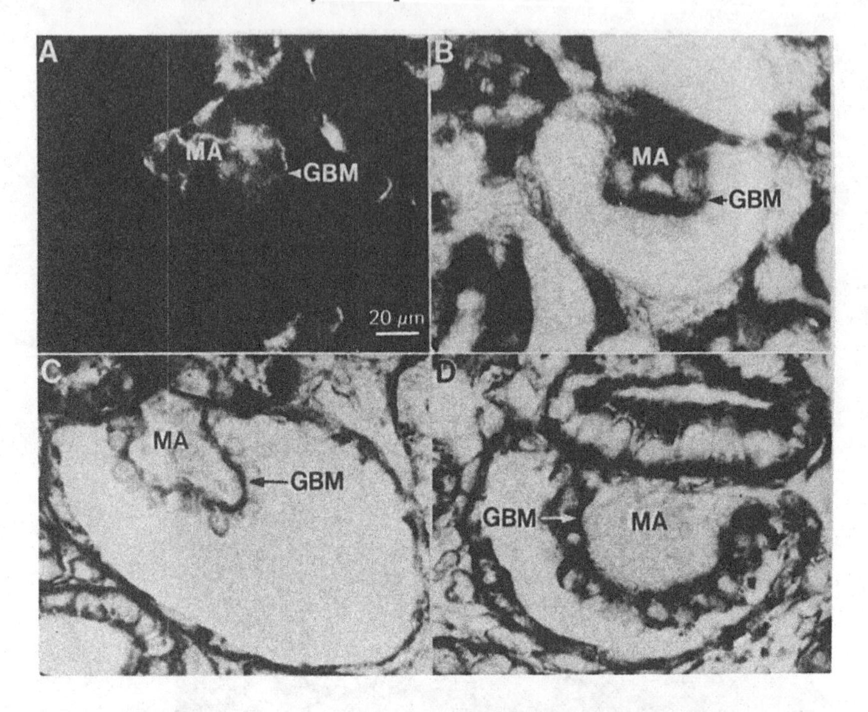

Fig. 5.6. Illustration of an analysis of the constituents of the glomerular basement membrane (GBM) and the mesangial area (MA) in mouse/chick chimaeric glomeruli (Sariola *et al.*, 1984*b*; courtesy of Dr H. Sariola). Immunofluorescence and peroxidase techniques after treatment with the following antisera. A. Monoclonal antibody against chick collagen type IV. B. Rabbit antibody against chick collagen type IV. C. Rabbit antibody against mouse collagen type IV. D. Affinity-purified antibody against mouse laminin.

glomeruli described earlier (Bonadio *et al.*, 1984). It may be suggested that the incomplete fusion of the two basement membranes is due to some molecular differences and mismatching (also reflected in the immunologically demonstrated species specificity).

The double-layered structure of the GBM has been beautifully visualized by Abrahamson (1985). The technique involved coupling of anti-laminin immunoglobulin G (IgG) to horseradish peroxidase (HRP) and injection of the probe into newborn rats. Peroxidase histochemistry on sections from the kidney cortex showed binding of the IgG–HRP complex to the basement membranes of early renal vesicles, S-shaped bodies and of the mature nephrons. At the capillary-loop stage, laminin was detected in the GBM as a double layer (Fig. 5.7).

Dual origin of the GBM was further supported by the observations of Abrahamson (1985) that application of the anti-laminin IgG–HRP complex to sections of newborn rat kidneys detected the label intracellularly

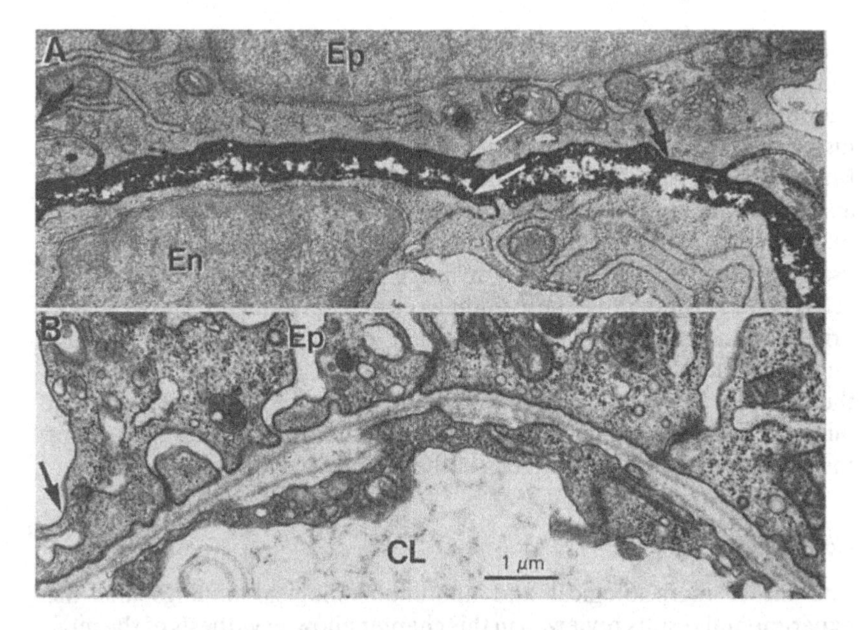

Fig. 5.7. Electron micrographs of the capillary wall of glomeruli at the capillary loop stage (Abrahamson, 1985; courtesy of Dr D. Abrahamson). Ep, glomerular epithelium; En, endothelium; CL, capillary lumen. A. Section from a newborn rat pulse-fixed with anti-laminin Immunoglobulin G–horseradish peroxidase. A double layer of peroxidase-positive material is seen within the basement membrane (arrows). B. Standard electron-microscopic technique with uninjected rats clearly demonstrates the double basement membrane (arrow).

both in the epithelial podocytes and the endothelial cells of the immature glomeruli.

The mesangial cells of the glomerulus have not yet been examined fully. They lie in close association with the capillary endothelium, but are not connected to their lumen. No specific markers have been reported for them (for a review, see Kreisberg & Karnovsky, 1983). The mesangial cells express a basement membrane-type matrix as well as fibronectin (Stenman & Vaheri, 1978; Oberley *et al.*, 1979; Courtoy *et al.*, 1980; Madri *et al.*, 1980; Ekblom *et al.*, 1981b).

The origin of the mesangial cells is still controversial. Suzuki (1959) and Vernier & Birch-Anderson (1962) concluded that the cells are formed *in situ* from the glomerular anlage, whereas Kazimierczak (1971) attributed their origin to mesenchymal cells that invade the glomerular crevice and become trapped between the endothelial loops. Bernstein *et al.* (1981) could not find mesangial cells in the avascular but otherwise well-developed glomerular bodies in the transfilter-induced mesenchymes. Hence, they regarded these cells as vascular derivatives. In the murine/avian chimaeric glomeruli described, cells were detected in association with the endothelium but not connected to the capillary lumen. These cells, mesangial by localization but otherwise indistinguishable from endothelial cells, showed characteristics of the avian host in immunohistology by expressing chicken collagen type IV and fibronectin. Mouse-specific anti-collagen type IV, anti-laminin and anti-fibronectin antibodies did not react with the mesangial area (Sariola *et al.*, 1984*a*, *b*, *c*). The avian origin of the mesangial cells could also be visualized by the use of the quail-specific antibody against endothelial (and hematopoietic) cells (Figs. 5.3 and 5.6). These findings clearly speak in favour of outside origin of the mesangial cells. They are either directly derived from migrating endothelial cells or their precursors co-migrate with endothelial cells. Final conclusions to confirm these findings based on light microscopy should come from electron-immunological examinations of the distribution of avian antigens in the chimaeric glomeruli.

Conclusions

The direct morphological and immunohistological findings and the experimental results reviewed in this chapter allow a synthesis of the most probable mechanism of vascularization of the nephron: a pair of vessels sprout from the dorsal aorta (Salama *et al.*, 1982) and invade the metanephric blastema in close association with the ingrowing ureteric bud. When the latter branches and induces pretubular aggregates around its tips, the migrating endothelial cells follow the border between these condensed areas and the fibronectin-expressing stroma. When the S-shaped bodies are formed, this path directs the endothelial cells into the first-formed lower crevice of the glomerular anlage. Here migration ceases, and the endothelial basement membrane fuses to that of the epithelial podocytes to create the final GBM. This spatially and temporally synchronized development of the mesenchymal and endothelial cell lineages is primarily programmed by the branching ureter and its inductive action. The latter determines the constitution and local changes in the ECM that ultimately trigger angiogenesis and guide the capillaries to their final location.

6
Concluding remarks

Control mechanisms

Figure 6.1, modified from previous schemes by Lehtonen & Saxén (1986*a*) and Saxén & Lehtonen (1986), summarizes much of our present knowledge of various molecular and structural events linked to the early, postinductory development of the secretory nephron. This process should be considered as a simple example of organogenesis singled out from a more complex process and consisting of a variety of developmental events not unique to the kidney: induction, proliferation, alternations in the intracellular and extracellular protein constituents, changes in cell shape and motility, and ultimately the assembly of cells into aggregates which gradually adopt organ- and tissue-specific formations with specific cellular phenotypes. These various processes have been documented in the previous chapters, but it is still not easy to find causal relationships within the general framework. Needless to say, such an analytical exercise will comprise – in addition to observations and facts – speculations and hypotheses.

The three main types of response of the nephric mesenchymal cells to an inductive stimulus are listed in the chart: documented events, their postulated and mostly unknown molecular basis, and their apparent morphogenetic consequences. The observed postinductory changes can be divided into three main types: (1) stimulation of the DNA synthesis (and proliferation) of the target cells, (2) disappearance (degradation) of the interstitial-type proteins from the extracellular matrix, and (3) enhanced (or neo-) synthesis of epithelial-type proteins of the extracellular matrix (ECM) and the cytoskeleton.

Proliferation of the induced cells is apparently stimulated and maintained by a dual control mechanism. The initial stimulation (12 to 24 h) depends on a persisting contact with the inductor and is probably emitted by a mitogen. Soluble growth factors are not required at this stage. After peak synthesis, inductor-dependence no longer exists, but the cells have become responsive to organismal growth factors (transferrin). How the inductor renders cells responsive to transferrin is not known. The most feasible explanation would be an induction of transferrin receptors, but

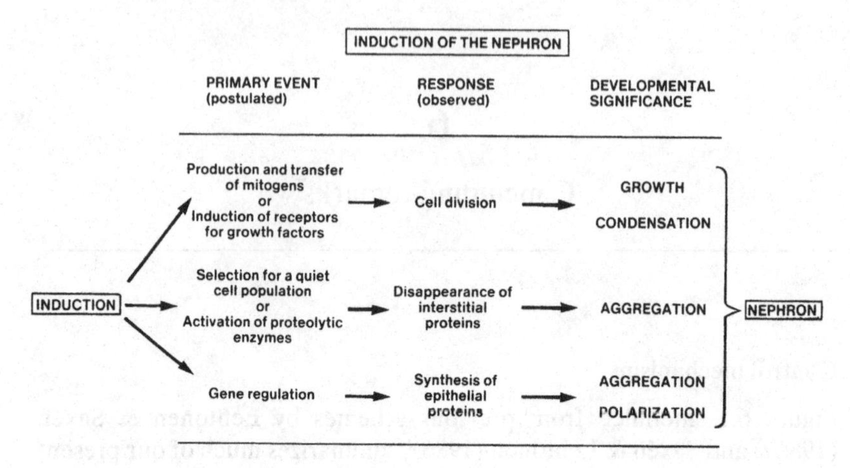

Fig. 6.1. Scheme of the postulated primary events, and their consequences, in the induction of the metanephric mesenchyme (for details, see the text) (Saxén & Lehtonen, 1986).

thus far efforts to document this have failed (Thesleff *et al.*, 1985).

Many of the consequences of a stimulated proliferation are apparent. Increased cell number is required for the growth of the anlage and probably also for the aggregation process itself. In addition, many examples are known, and well documented, where expression of a new cellular phenotype requires mitosis. We do not know whether the observed stimulation of proliferation in the nephric mesenchyme is selective, i.e. whether the mitogen acts upon a selected, predetermined cell lineage subsequently expressing an epithelial phenotype. The possibility is not fully excluded that the seemingly homogeneous cell population within the nephric mesenchyme contains predetermined subpopulations with restricted developmental options and selective sensitivity to mitotic stimulation.

Disappearance of the interstitial proteins occurs concomitantly with the initial stimulation of DNA synthesis at 12 to 24 h. Two possible mechanisms should be considered; either the rapid proliferation of induced cells leads to a cell population devoid of synthesis of the interstitial marker proteins or induction leads to an activation of proteolytic enzymes degrading the interstitial proteins. The 12-h period during which the compounds disappear might be sufficient for one cell cycle to produce a second, quiescent generation of cells, but it appears to be rather short for the non-induced disappearance of the collagenous and non-collagenous compounds of the matrix. Thus, disappearance by digestion of activated proteolytic enzymes seems a more plausible mechanism, and this would also explain the reappearance of fibronectin in monolayer cultures after the removal of the inductor. Despite repeated efforts, we have failed to show such enzymic activity in the cultured mesenchymal explants or any split products of the proteins in the culture medium (unpublished work).

It is not difficult to speculate upon the consequences of changes in the constitution of the ECM. Diminution of the ECM should lead to increased cell density (condensation) and decreased cell motility – both observed in the induced mesenchyme. Restricted intercellular spaces further facilitate cell–cell interactions within the aggregates, leading to cell organization and subsequent adhesion (below). An additional consequence of the orderly changes in the ECM was dealt with in Chapter 5, where the guided migration of the kidney capillaries was discussed. The paths which the endothelial cells follow might well be created by the changing ECM.

Enhanced synthesis of the epithelial proteins was detected and measured throughout the early stages of nephron formation from the first precondensation phase to the segmentation of the tubules. These proteins naturally vary in their functions and consequences and include components of the basement membrane and the lateral cell membranes as well as proteins of the epithelial cytoskeleton and those with enzymic activity. As long as the signal substances transmitting the message from the inductor to the target cells are unknown, the mechanism for an enhanced or new protein synthesis should be left open. We cannot yet predict whether the inductor activates several structural genes and their regulators or whether a sequential gene activation is triggered by a single message. The technology and probes now available should unravel this fundamental problem in the not too distant future.

Consequences of the enhanced or new synthesis of the epithelial proteins are, indeed, numerous, and here we may deal only with those with an apparent morphogenetic significance. Cells brought together initially by the changes in the ECM are soon aggregated into colonies of cells with increased intercellular adhesiveness. This is most probably mediated by the synthesis of new proteins with adhesive properties or an enhanced production of such 'ligands'. Several proteins with known adhesive properties have been described during the initial aggregation phase in the ECM of the kidney mesenchyme. These include laminin (Ekblom *et al.*, 1980*a*), neural cell adhesion molecule (N-CAM) (Thiery *et al.*, 1982*a*) and uvomorulin (Vestweber *et al.*, 1985). It remains open which, if any of these is the key molecule.

The primary aggregate soon becomes stabilized by the attachment of cells to the basement membrane and by the development of lateral membrane junctions. The basement membrane built up by a set of compounds synthesized by the induced and aggregated cells certainly maintains the epithelial assembly of cells in the newly formed pretubular aggregates, but its polarizing and epithelializing effects are matter for discussion – in fact the peripheral accumulation of the basement membrane material itself is an expression of polarization. Following the classic ideas of Gustafsson & Wolpert (1963), polarization of the pre-epithelial mesenchymal cells could be traced back to the early aggregation phase. An increased intercellular adhesiveness within the aggregates should, accordingly, lead to alternations in the cell shape and initiation of polarization. Moreover, I have already cited many examples where certain epithelial features can be achieved in experimental conditions which exclude the effect of a basement membrane. These include the onset of enzymic activity of pancreatic exocrine cells in monolayer cultures (Spooner *et al.*, 1977), the appearance of cytokeratin in the induced kidney mesenchyme in monolayer cultures (Lehtonen *et al.*, 1985), and induction of skin appendages with region-specific keratin patterns in conditions preventing the formation of the basement membrane (Peterson & Grainger, 1985).

Formation of the basement membrane is followed by the establishment of cytokeratin-type intermediate filaments apparently contributing to the establishment of the epithelial features of the tubule cells. Many models have been presented for the 'morphogenetic axis' extracellular matrix – trans-membrane proteins – cytoskeleton (e.g. Hay, 1984). Observations in the kidney system can be fitted in well with those models.

References

Abrahamson, D. R. (1985). Origin of the glomerular basement membrane visualized after *in vivo* labeling of laminin in newborn rat kidneys. *J. Cell Biol.* **100**, 1988–2000.

Alescio, T. & Cassini, A. (1962). Induction *in vitro* of tracheal buds by pulmonary mesenchyme grafted on tracheal epithelium. *J. Exp. Zool.* **150**, 83–94.

Aoki, A. (1966). Development of the human renal glomerulus. I. Differentiation of the filtering membrane. *Anat. Rec.* **155**, 339–52.

Armstrong, P. B. & Armstrong, M. T. (1973). Are cells in solid tissues immobile? Mesonephric mesenchyme studied *in vitro*. *Devel. Biol.* **35**. 187–209.

Atterbury, R. R. (1923). Development of the metanephric anlage of chick in allantoic grafts. *Am. J. Anat.* **31**, 409–36.

Auer, J. (1947). Bilateral renal agenesia. *Anat. Rec.* **97**, 283–92.

Auerbach, R. (1972). The use of tumors in the analysis of inductive tissue interactions. *Devel. Biol.* **28**, 304–9.

Auerbach, R. (1977). Toward a developmental theory of immunity: selective differentiation of teratoma cells. In *Cell and Tissue Interations*, ed. J. W. Lash & M. M. Burger, pp. 47–55. Raven Press, New York.

Auerbach, R. & Grobstein, C. (1958). Inductive interaction of embryonic tissues after dissociation and reaggregation. *Exp. Cell Res.* **15**, 384–97.

Auerbach, R., Kubai, L. & Sidky, Y. (1976). Angiogenesis induction by tumors, embryonic tissues and lymphocytes. *Cancer Res.* **36**, 3435–40.

Avner, E. D., Ellis, D., Temple, T. & Jaffe, R. (1982). Metanephric development in serum-free organ culture. *In Vitro*, **18**, 675–82.

Avner, E. D., Jaffe, R., Temple, T., Ellis, D. & Chung, A. E. (1983). Development of renal basement membrane glycoproteins in metanephric organ culture. *Lab. Invest.* **48**, 263–8.

Barakat, T. I. & Harrison, R. G. (1971). The capacity of fetal and neonatal renal tissues to regenerate and differentiate in a heterotopic allogeneic subcutaneous tissue site in the rat. *J. Anat.* **110**, 393–407.

Barnes, D. & Sato, G. (1980). Serum-free cell culture, a unifying approach. *Cell* **22**, 649–55.

Bernfield, M. R., Banerjee, S. D. & Cohn, R. H. (1972). Dependence of salivary epithelial morphology and branching morphogenesis upon acid mucopolysaccharide-protein (proteoglycan) at the epithelial surface. *J. Cell Biol.* **52**, 674–89.

Bernfield, M., Banerjee, S. D., Koda, J. E. & Rapraeger, A. C. (1984*a*). Remodelling of basement membrane as a mechanism of morphogenetic tissue interaction. In *The Role of Extracellular Matrix in Development*, ed. R. L. Trelstad, pp. 545–72. Alan R. Liss, New York.

Bernfield, M., Banerjee, S. D., Koda, J. E. & Rapraeger, A. C. (1984*b*). Remodelling of the basement membrane: morphogenesis and maturation. In *Basement Membranes and Cell Movement*, ed. R. Porter & J. Wheland, pp. 179–96. Pitman, London.

Bernstein, J. (1978). Morphologic development of the metanephric tubule. In *Proceedings of the VIIth International Congress on Nephrology*, ed. M. Bergeron, pp. 249–54. S. Karger, Basel.
Bernstein, J., Cheng, F. & Roszka, J. (1981). Glomerular differentiation in metanephric culture. *Lab. Invest.* **45**, 183–90.
Berton, J. P. (1965). Anatomie vasculaire de mésonéphros chez certains mammifères. I. Les mésonéphros de l'embryon de porc. *C. R. Assoc. Anat.* **124**, 272–90.
Biggers, J. D., Gwatkin, R. B. L. & Heynes, S. (1961). The growth of avian and mammalian tibiae on a relatively simple chemically defined medium. *Exp. Cell Res.* **25**, 41–58.
Bijtel, J. H. (1948). Deviation experimental de l'uretère primaire par la greffe de materiel cloaca chez le *Triton. Experientia* **4**, 1–5.
Bijtel, J. H. (1968). Experimental data concerning the development of the pronephric duct in axolotl. *Acta Morphol.* **7**, 91.
Birchmeier, W., Imhof, B. A., Goodman, S. L. & Vollmers, H. P. (1985). Functional monoclonal antibodies in the study of cell adhesion, tumor cell growth, and metastasis. In *Gene Expression during Normal and Malignant Differentiation*, ed. L. C. Andersson, C. G. Gahmberg & P. Ekblom, pp. 33–55. Academic Press, New York.
Bishop-Calame, S. (1965*a*). Sur le comportement en greffes chorio-allantoidiennes, de l'uretère de l'embryon de poulet associé à mésenchymes hétérologues. *C. R. Acad. Sci. Paris* **260**, 281–3.
Bishop-Calame, S. (1965*b*). Etude d'associations hétérologues de l'uretère et de différents mésenchymes de l'embryon de poulet, par la technique des greffes chorio-allantoidiennes. *J. Embryol. Exp. Morphol.* **14**, 247–53.
Bishop-Calame, S. (1966). Etude expérimentale de l'organogenèse du système urogénital de l'embryon de poulet. *Arch. Nat. Microscop. Exptl.* **55**, 215–309.
Bonadio, J. F., Sage, H., Cheng, F., Bernstein, J. & Striker, G. E. (1984). Localization of collagen types IV and V, laminin, and heparan sulfate proteoglycan to the basal lamina of kidney epithelial cells in transfilter metanephric culture. *Am. J. Pathol.* **116**, 287–96.
Borghese, E. (1950). Explanation experiments on the influence of the connective tissue capsule on the development of the epithelial part of the submandibular gland of *Mus musculus. J. Anat.* **84**, 303–18.
Bosshard, H. J. (1971). Experimentelle analyse zur Entwicklung der chimärischen Vorniere bei *Triturus* und *Bombina. Roux' Arch.* **168**, 282–303.
Boyden, E. A. (1927). Experimental obstruction of the mesonephric ducts. *Proc. Soc. Exp. Biol. Med.* **24**, 572–6.
Boyden, E. A. (1932). Congenital absence of the kidney: an interpretation based on a 10-mm human embryo exhibiting unilateral renal agenesis. *Anat. Rec.* **52**, 325–39.
Brauer, A. (1902). Beiträge zur Kenntniss der Entwicklung und Anatomie der Gymnophionen. III. Die Entwicklung der Excretionsorgane. *Zool. Jb. Anat. Ontog. Tiere* **16**, 1–176.
Bremer, J. L. (1915). The mesonephric corpuscle of the sheep, cow, and deer. *Anat. Rec.* **10**, 1–7.
Burger, M. M. (1974). Mechanisms of cell–cell recognition: some comparisons between lower organisms and vertebrates. In *Cell Interactions in Differentiation*, eds. M. Karkinen-Jääskeläinen, L. Saxén & L. Weiss, pp. 357–76. Academic Press, London.
Burns, R. K. (1938). Development of the mesonephros in Amblystoma after early extirpation of the duct. *Proc. Soc. Exp. Biol. Med.* **39**, 111–13.
Burns, R. K. (1955). Urogenital system. In *Analysis of Development*, ed. B. J. Willier, P. Weiss & V. Hamburger, pp. 462–91. W. B. Saunders, Philadelphia.
Cade-Treyer, D. (1972). Isolation of pure fractions of viable calf kidney tubules and

glomeruli. *In vitro* culture, immunochemical and esterase zymogram analysis. *Ann. Inst. Pasteur* **122**, 263–82.
Cade-Treyer, D. & Tsuji, S. (1975). *In vitro* culture of the proximal tubule of the bovine nephron. *Cell Tissue Res.* **163**, 5–28.
Calame, S. (1959). Sur les relations entre le canal de Wolff et le développement du mésonephros et de la gonade chez l'embryon de l'oiseau. *C. R. Acad. Sci. Paris* **248**, 3033–5.
Calame, S. (1961). Le rôle des composants épithélial et mésenchymateoux du métanéphros, d'après les résultats de la culture *in vitro*. *Arch. Anat. Microsc. Morphol. Exp.* **50**, 299–308.
Calame, S. (1962). Contribution expérimentale à l'étude du développement du système urogénital de l'embryon d'oiseau. *Arch. Anat. Strasb.* **94**, 45–65.
Cambar, R. (1948). Recherches expérimentales sur les facteurs de la morphogénèse du mésonéphros chez les amphibiens anoures. *Bull. Biol. Fr. Belg.* **82**, 214–85.
Cambar, R. (1952*a*). L'uretère primaire de la grenouille se développe, en direction postérieure, à partir d'un blastème voisin du pronéphros: démonstration expérimentale. *C. R. Soc. Biol.* **146**, 77–80.
Cambar, R. (1952*b*). Essai expérimental de développement de l'uretère primaire des amphibiens anoures en milieu atypique. *C. R. Soc. Biol.* **146**, 453–5.
Carlin, B., Jaffe, R., Bender, B. & Chung, A. E. (1981). Entactin, a novel basal lamina-associated sulfated glycoprotein. *J. Biol. Chem.* **256**, 5209–14.
Carpenter, K. L. & Turpen, J. B. (1979). Experimental studies on hemopoiesis in the pronephros of *Rana pipiens*. *Differentiation* **14**, 167–74.
Cooper, G. W. (1965). Induction of somite chondrogenesis by cartilage and notochord: a correlation between inductive activity and specific stages of cytodifferentiation. *Devel. Biol.* **12**, 185–212.
Courtoy, P. J., Kanwar, Y. S., Hynes, R. O. & Farquhar, M. G. (1980). Fibronectin localization in the rat glomerulus. *J. Cell Biol.* **87**, 691–6.
Courtoy, P. J., Timpl, R. & Farquhar, M. G. (1982). Comparative distribution of laminin, type IV collagen, and fibronectin in the rat kidney cortex. *J. Histochem. Cytochem.* **30**, 874–86.
Crocker, J. F. S. (1973). Human embryonic kidneys in organ culture: abnormalities of development induced by decreased potassium. *Science* **181**, 1178–9.
Crocker, J. F. S. & Vernier, R. L. (1970). Fetal kidney in organ culture: abnormalities of development induced by decreased amounts of potassium. *Science* **169**, 485–7.
Croisille, Y. (1969). Détection et localization de constituants spécifiques de rein à des stades précoces de la tubulogenèse chez les embryos de poulet et de caille. *Bull. Biol. Fr. Belg.* **103**, 339–73.
Croisille, Y. (1970). Appearance and disappearance of organ-specific components during kidney tubulogenesis in chick and quail embryos. In *Protides of the Biological Fluids*, ed. H. Peeters, pp. 79–85. Pergamon Press, Oxford.
Croisille, Y. (1976). On some recent contributions to the study of kidney tubulogenesis in mammals and birds. In *Tests of Teratogenicity in Vitro*, ed. J. D. Ebert & M. Marois, pp. 149–70. North-Holland Publ. Co., Amsterdam.
Croisille, Y., Gumpel-Pinot, M. & Martin, C. (1971). Sur l'organogenèse du mésonépros chez les oiseaux. Etude immunohistologique du tubule urinaire chez l'embryon de poulet. *C. R. Acad. Sci. Paris* **272**, 629–31.
Croisille, Y., Gumpel-Pinot, M. & Martin, C. (1974). Immunochimie et développement: organigenèse du rein chez les oiseaux. *Ann. Biol.* 13, 325–40.
Croisille, Y., Gumpel-Pinot, M. & Martin, C. (1976). Embryologie expérimentale. La différenciation des tubes sécréteurs du rein chez les oiseaux: effects des inducteurs hétérogènes. *C. R. Acad. Sci. Paris Ser. D.* **282**, 1987–90.
Cunha, G. R. (1975). The dual origin of vaginal epithelium. *Am. J. Anat.* **143**, 387–92.

Cunha, G. R. (1985). Mesenchymal-epithelial interactions during androgen-induced development of the prostate. In *Developmental Mechanisms: Normal and Abnormal*, ed. J. W. Lash & L. Saxén, pp. 15–24. Alan R. Liss, Inc., New York.

Cunha, G. R., Chung, L. W. K., Shannon, J. M., Taguchi, O. & Fuju, H. (1983). Hormone-induced morphogenesis and growth: role of mesenchymal–epithelial interactions. *Rec. Prog. Horm. Res.* **39**, 560–98.

Cunha, G. R., Bigsby, R. M., Cooke, P. S. & Sugimura, Y. (1985). Stromal–epithelial interactions in adult organs. *Cell Different.* **17**, 137–48.

Darmady, E. M. & Stranack, F. (1957). Microdissection of the nephron in disease. *Brit. Med. Bull.* **13**, 21–6.

Davies, J. (1950). The pronephros and the early development of the mesonephros in the duck. *J. Anat.* **84**, 95–103.

Dawnay, A., McLean, C. & Cattel, W. R. (1980). The development of radioimmunoassay for Tamm–Horsfall glycoprotein in serum. *Biochem. J.* **185**, 679–87.

De Martino, C. & Zamboni, L. (1966). A morphologic study of the mesonephros of the human embryo. *J. Ultrastr. Res.* **16**, 399–427.

Deuchar, E. M. (1970). Neural induction and differentiation with minimal numbers of cells. *Devel. Biol.* **22**, 185–99.

Dørup, J. & Maunsbach, A. B. (1982). The ultrastructural development of distal nephron segments in the human fetal kidney. *Anat. Embryol.* **164**, 19–41.

Du Bois, A. M. (1969). The embryonic kidney. In *The Kidney*, ed. C. Rouiller & A. F. Muller, vol. I, pp. 1–59. Academic Press, New York.

Edelman, G. M. (1983). Cell adhesion molecules. *Science* **219**, 450–7.

Edelman, G. M. (1985). Expression of cell adhesion molecules during embryogenesis and regeneration. *Exp. Cell. Res.* **161**, 1–16.

Edwards, J. G. (1951). The development of the efferent arteriole in human metanephros. *Anat. Rec.* **109**, 495–502.

Ekblom, P. (1981*a*). Formation of basement membranes in the embryonic kidney: an immunohistological study. *J. Cell. Biol.* **91**, 1–10.

Ekblom, P. (1981*b*). Determination and differentiation of the nephron. *Med. Biol.* **59**, 139–60.

Ekblom, P. (1984). Basement membrane proteins and growth factors in kidney differentiation. In *Role of Extracellular Matrix in Development*, ed. R. L. Trelstad, pp. 173–206. Alan R. Liss, New York.

Ekblom, P. & Thesleff, I. (1985). Control of kidney differentiation by soluble factors secreted by the embryonic liver and yolk sac. *Devel. Biol.* **101**, 29–38.

Ekblom, P., Lash, J. W., Lehtonen, E., Nordling, S. & Saxén, L. (1979*a*). Inhibition of morphogenetic cell interactions by 6-diazo-5-oxo-norleucine (DON). *Exp. Cell Res.* **121**, 121–6.

Ekblom, P., Nordling, S., Saxén, L., Rasilo, M.-L. & Renkonen, O. (1979*b*). Cell interactions leading to kidney tubule determination are tunicamycin sensitive. *Cell Different.* **8**, 347–52.

Ekblom, P., Alitalo, K., Vaheri, A., Timpl, R. & Saxén, L. (1980*a*). Induction of a basement membrane glycoprotein in embryonic kidney: possible role of laminin in morphogenesis. *Proc. Nat. Acad. Sci. USA* **77**, 485–9.

Ekblom, P., Miettinen, A. & Saxén, L. (1980*b*). Induction of brush border antigens of the proximal tubule in the developing kidney. *Devel. Biol.* **74**, 263–74.

Ekblom, P., Miettinen, A., Virtanen, I., Wahlström, T., Dawnay, A. & Saxén, L. (1981*a*). *In vitro* segregation of the metanephric nephron. *Devel. Biol.* **84**, 88–95.

Ekblom, P., Lehtonen, E., Saxén, L. & Timpl, R. (1981*b*). Shift in collagen types as an early response to induction of the metanephric mesenchyme. *J. Cell Biol.* **89**, 276–83.

Ekblom, P., Thesleff, I., Miettinen, A. & Saxén, L. (1981*c*). Organogenesis in a defined medium supplemented with transferrin. *Cell Different.* **10**, 281–8.
Ekblom, P., Sariola, H., Karkinen-Jääskeläinen, M. & Saxén, L. (1982). The origin of the glomerular endothelium. *Cell Different.* **11**, 35–9.
Ekblom, P., Thesleff, I., Saxén, L., Miettinen, A. & Timpl, R. (1983). Transferrin as a fetal growth factor: acquisition of responsiveness related to embryonic induction. *Proc. Nat. Acad. Sci. USA* **80**, 2651–5.
Emura, M. & Tanaka, T. (1972). Development of endothelial and erythroid cells in mouse metanephric mesenchyme cultured with fetal liver. *Devel. Growth Different.* **14**, 237–46.
Etheridge, A. L. (1968). Determination of the mesonephric kidney. *J. Exp. Zool.* **169**, 357–70.
Evan, A. P., Gattone, V. H., II & Blomgren, P. M. (1984). Application of scanning electron microscopy to kidney development and nephron maturation. *Scanning Electron Microsc.* **I**, 455–73.
Fales, D. E. (1935). Experiments on the development of the pronephros of *Amblystoma punctatum. J. Exp. Zool.* **72**, 147–73.
Farquhar, M. G. (1981). The glomerular basement membrane. A selective macromolecular filter. In *Cell Biology of Extracellular Matrix*, ed. E. D. Hay, pp. 335–78. Plenum Press, New York.
Feinberg, R. N. & Beebe, D. C. (1983). Hyaluronate in vasculogenesis. *Science* **220**, 1177–9.
Fell, P. E. & Grobstein, C. (1968). The influence of extra-epithelial factors on the growth of embryonic mouse pancreatic epithelium. *Exp. Cell Res.* **53**, 301–4.
Field, H. H. (1891). The development of the pronephros and segmental duct in amphibia. *Bull. Museum Comp. Zool. Harvard* **21**, 201–340.
Fleischmajer, R. & Billingham, R. E. (eds.) (1968). *Epithelial–Mesenchymal Interactions*, pp. 1–326. Williams & Wilkins, Baltimore.
Foidart, J. B., Dechenne, C. A., Mahieu, P., Creutz, C. E. & de Mey, J. (1979). Tissue culture of normal rat glomeruli. *Invest. Cell Pathol.* **2**, 15–26.
Folkman, J. & Cotran, R. (1976). Relation of vascular proliferation to tumor growth. *Int. Rev. Exp. Pathol.* **76**, 207–48.
Fox, H. (1956). Compensation in a remaining pronephros of *Triturus* after unilateral pronephrectomy. *J. Embryol. Exp. Morphol.* **4**, 139–51.
Fox. H. (1963). The amphibian pronephros. *Quart. Rev. Biol.* **38**, 1–25.
Franke, W. W., Schmid, E., Schiller, D. L., Winter, S., Jarasch, E.-D., Moll, R., Denk, H., Jackson, B. W. & Illmensee, K. (1982). Differentiation-related patterns of expression of proteins of intermediate-sized filaments in tissue and cultured cells. *Cold Spring Harbor Symp. Quant. Biol.* **46**, 431–53.
Fraser, E. A. (1950). The development of the vertebrate excretory system. *Biol. Rev. Camb. Phil. Soc.* **25**, 159–87.
Friebová-Zemanová, Z. & Concharevskaya, O. A. (1982). Formation of the chick mesonepros. 5. Spatial distribution of the nephron populations. *Anat. Embryol.* **165**, 125–39.
Gluecksohn-Schoenheimer, S. (1949). Causal analysis of mouse development by the study of mutational factors. *Growth Symposium* **9**, Suppl., 163–76.
Golgi, C. (1889). Annotazioni intorno all-istologia dei reni dell'uomo e di altri mammiferi e sull'istogenesi dei canalicoli uriniferi. *Rendic. Accad. Lincei* **14**, 286–320.
Goodrich, E. S. (1930). *Studies of the Structure and Development of the Vertebrates.* Macmillan Co., London.
Gospodarowicz, D. & Tauber, J. P. (1980). Growth factors and the extracellular matrix. *Endocrinol. Rev.* **1**, 201–27.

Gospodarowicz, D., Greenburg, G., Foidart, J.-M. & Savion, N. (1981). The production and localization of laminin in cultured vascular and corneal endothelial cells. *J. Cell. Physiol.* **107**, 171–83.

Gospodarowicz, D., Cohen, D. C. & Massoglia, S. L. (1983). Stimulation of the proliferation of the Madin–Darby canine kidney (MDCK) epithelial cell line by high-density lipoproteins and their induction of 3-hydroxy-3-methylglutaryl coenzyme A reductase activity. *J. Cell Physiol* **117**, 76–90.

Gospodarowicz, D., Lepine, J., Massoglia, S. & Wood, I. (1984). Comparison of the ability of basement membranes produced by corneal endothelial and mouse-derived endodermal PF-HR-9 cells to support the proliferation and differentiation of bovine kidney tubule epithelial cells *in vitro*. *J. Cell Biol.* **99**, 947–61.

Gossens, C. L. & Unsworth, B. R. (1972). Evidence for a two-step mechanism operating during *in vitro* mouse kidney tubulogenesis. *J. Embryol. Exp. Morphol.* **28**, 615–31.

Grobstein, C. (1953*a*). Epithelio-mesenchymal specificity in the morphogenesis of mouse submandibular rudiments *in vitro*. *J. Exp. Zool.* **124**, 383–414.

Grobstein, C. (1953*b*). Morphogenetic interaction between embryonic mouse tissues separated by a membrane filter. *Nature (Lond.)* **172**, 869–71.

Grobstein, C. (1953*c*). Inductive epithelio-mesenchymal interaction in cultured organ rudiments of the mouse. *Science* **118**, 52–5.

Grobstein, C. (1955*a*). Inductive interaction in the development of the mouse metanephros. *J. Exp. Zool.* **130**, 319–40.

Grobstein, C. (1955*b*). Tissue interaction in the morphogenesis of mouse embryonic rudiments *in vitro*. In *Aspects of Synthesis and Order in Growth*, ed. D. Rudnick, pp. 233–56. Princeton University Press, Princeton, NJ.

Grobstein, C. (1956*a*). Trans-filter induction of tubules in mouse metanephrogenic mesenchyme. *Exp. Cell Res.* **10**, 424–40.

Grobstein, C. (1956*b*). Inductive tissue interaction in development. *Adv. Cancer Res.* **4**, 187–236.

Grobstein, C. (1957). Some transmission characteristics of the tubule-inducing influence on mouse metanephrogenic mesenchyme. *Exp. Cell Res.* **13**, 575–87.

Grobstein, C. (1961). Cell contact in relation to embryonic induction. *Exp. Cell Res.* Suppl. **8**, 234–45.

Grobstein, C. (1962). Interactive processes in cytodifferentiation. *J. Cell Comp. Physiol.* **60**, Suppl. 1, 35–48.

Grobstein, C. (1967). Mechanisms of organogenetic tissue interaction. *Nat. Cancer Inst. Monogr.* **26**, 279–99.

Grobstein, C. & Dalton, A. J. (1957). Kidney tubule induction in mouse metanephrogenic mesenchyme without cytoplasmic contact. *J. Exp. Zool.* **135**, 57–73.

Grobstein, C. & Parker, G. (1958). Epithelial tubule formation by mouse metanephrogenic mesenchyme transplanted *in vivo*. *J. Nat. Cancer Inst.* **2**, 107–19.

Gruenwald, P. (1937). Zur Entwicklungsmechanik der Urogenital-systems beim Huhn, *Roux' Arch.* **136**, 786–813.

Gruenwald, P. (1942). Experiments on the distribution and activation of the nephrogenic potency in the embryonic mesenchyme. *Physiol. Zool.* **15**, 396–409.

Gruenwald, P. (1943). Stimulation of nephrogenic tissues by normal and abnormal inductors. *Anat. Rec.* **86**, 321–35.

Gruenwald, P. (1952). Development of the excretory system. *Ann. N.Y. Acad. Sci.* **55**, 142–6.

Gullino, P. M. (1981). Tissue growth factors. In *Handbook of Experimental Pharmacology*, vol. 57, ed. R. Baserga, pp. 427–49. Springer Verlag, New York.

Gumpel-Pinot, M., Martin, C. & Croisille, Y. (1971). Sur l'organogenèse du

mésonéphros chez les oiseaux: réalisation *in vitro* de mésonéphros chimères caille-poulet. *C. R. Acad. Sci. Paris* **272**, 737–9.
Gustafsson, T. & Wolpert, L. (1963). The cellular basis of morphogenesis and sea urchin development. *Int. Rev. Cytol.* **15**, 139–214.
Haffen, K. (1951). Contribution á l'étude de la régression morphologique et histologique du mésonéphros de l'embryon de poulet. *C.R. Séanc. Soc. Biol.* **145**, 755–9.
Hall, B. V. & Roth, E. (1957). Preliminary studies on the development and differentiation of cells and structures of the renal corpuscle. In *Proceedings of the Stockholm Conference on Electron Microscopy*, ed. F. S. Sjöstrand & J. Rhodin, pp. 176–9. Academic Press, New York.
Hamilton, H. (1952). *Lillie's Development of the Chick*. Henry Holt, New York.
Hassell, J. R., Robey, P. G., Barrach, H.-J., Wilczek, J., Rennard, S. I. & Martin, G. R. (1980). Isolation of a heparan sulfate-containing proteoglycan from basement membrane. *Proc. Nat. Acad. Sci. USA* **77**, 4494–8.
Hay, E. D. (1983). Cell and extracellular matrix: their organization and mutual dependence. In *Spatial Organization of Eukaryote Cells*, ed. J. R. McIntosh, pp. 509–48. Alan R. Liss, New York.
Hay, E. D. (1984). Cell-matrix interaction in the embryo: cell shape, cell surface, cell skeletons, and their role in differentiation. In *The Role of Extracellular Matrix in Development*, ed. R. L. Trelstad, pp. 1–31. Alan R. Liss, New York.
Herring, P. (1900). The development of the Malpighian bodies and its relation to pathologic changes which occur in them. *J. Pathol. Bacteriol.* **6**, 459–96.
Hilfer, S. R. & Hilfer, E. S. (1983). Computer simulation of organogenesis: an approach to the analysis of shape changes in epithelial organs. *Devel. Biol.* **97**, 444–53.
Hiller, S. (1931). Etude expérimentale sur la structure et la fonction du pronéphros dans la parabiose. *C. R. Assoc. Anat.* **26**, 267–80.
Hoar, R. M. & Monie, I. W. (1981). Comparative development of specific organ systems. In *Developmental Toxicology*, ed. C. A. Kimmel & J. Buelke-Sam, pp. 13–33. Raven Press, New York.
Holtfreter, J. (1933). Der Einfluss von Wirtsalter und verschiedenen Organbezirken auf die Differenzierung von angelagertem Gastrulaektoderm. *Roux' Arch.* **127**, 619–775.
Holtfreter, J. (1939). Gewebeaffinität, ein Mittel der embryonalen Formbildung. *Arch. exp. Zellf.* **23**, 169–209.
Holtfreter, J. (1944). Experimental studies on the development of the pronephros. *Rev. Can. Biol.* **3**, 220–50.
Holthöfer, H. & Virtanen, I. (1986). Glycosylation of developing human glomeruli: lectin binding sites during cell induction and maturation. *J. Histochem. Cytochem.* **34**, (in press).
Holthöfer, H., Miettinen, A., Lehto, V.-P., Lehtonen, E. & Virtanen, I. (1984). Expression of vimentin and cytokeratin types of intermediate filament proteins in developing and adult human kidneys. *Lab. Invest.* **50**, 552–9.
Holtzer, H., Bennett, G. S., Tapscott, S. J., Croop, J. M. & Toyama, Y. (1982). Intermediate-size filaments: changes in synthesis and distribution in cells of the myogenic and neurogenic lineages. *Cold Spring Harbor Symp. Quant. Biol.* **46**, 317–29.
Howland, R. B. (1921). Experiments on the effects of removal of the pronephros of *Amblystoma punctatum*. *J. Exp. Zool.* **32**, 355–95.
Hoyer, J. R., Resnick, J. S., Michael, A. D. & Vernier, R. L. (1974). Ontogeny of Tamm–Horsfall urinary glycoprotein. *Lab. Invest.* **30**, 757–61.
Huang, T. W. (1979). Basal lamina heterogeneity in the glomerular capillary tufts of human kidneys. *J. Exp. Med.* **149**, 1450–9.

Huber, C. (1905). On the development and shape of uriniferous tubules of certain higher mammals. *Am. J. Anat.* Suppl. **4**, 1–98.

Humphrey, R. R. (1928). The developmental potencies of the intermediate mesoderm of *Amblystoma* when transplanted into ventrolateral sites in other embryos: the primordial germ cells of such grafts and their role in the development of a gonad. *Anat. Rec.* **40**, 67–101.

Jacob. H. J., Jacob, M. & Christ, B. (1977). Die Ultrastruktur der externen Glomerula. Ein Beitrag zur Nierenentwicklung bei Hühnerembryonen. *Verh. Anat. Ges.* **71**, 909–12.

Jaffe, E. A., Minick, R., Adelman, B., Becker, C. D. & Nachman, R. (1976). Synthesis of basement membrane collagen by cultured human endothelial cells. *J. Exp. Med.* **144**, 209–25.

Jainchill, J., Saxén, L. & Vainio, T. (1964). Studies on kidney tubulogenesis. I. The effect of actinomycin D on tubulogenesis *in vitro*. *J Embryol. Exp. Morphol.* **12**, 597–607.

Jokelainen, P. (1963). An electron microscopic study of the early development of the rat metanephric nephron. *Acta Anat.* Suppl. **47**, 1–73.

Karkinen-Jääskeläinen, M. (1978). Transfilter lens induction in avian embryo. *Differentiation* **12**, 31–7.

Kazimierczak, J. (1965). Development of the glomerulus and juxtaglomerular apparatus. *Acta Pathol. Microbiol. Scand.* **65**, 318–20.

Kazimierczak, J. (1970). Histochemical observations of the developing glomerulus and juxtaglomerular apparatus. *Acta Pathol. Microbiol. Scand. ser.* A Suppl. **78**, 401–13.

Kazimierczak, J. (1971). Development of renal corpuscle and the juxtaglomerular apparatus: a light and electron microscopic study. *Acta Pathol. Microbiol. Scand. ser.* A Suppl. **218**, 1–16.

Kefalides, N. A., Alper, R. & Clark, C. C. (1979). Biochemistry and metabolism of basement membranes. *Int. Rev. Cytol.* **61**, 167–228.

Killen, P. D. & Striker, G. F. (1979). Human glomerular visceral epithelium synthesize a basal lamina collagen *in vitro*. *Proc. Nat. Acad. Sci. USA* **76**, 3518–22.

Kiremidjian, L. & Kopac, M. J. (1972). Changes in cell adhesiveness associated with the development of *Rana pipiens* pronephros. *Devel. Biol.* **27**, 116–30.

Koskimies, O. (1967*a*). Studies on kidney tubulogenesis. X. The effect of actinomycin D on the development of the lactate dehydrogenase isozyme pattern during tubule formation *in vitro*. *Exp. Cell Res.* **46**, 541–52.

Koskimies, O. (1967*b*). Cytodifferentiation of mouse kidney mesenchyme *in vitro*: developmental patterns of lactate dehydrogenase isozymes during tubule formation. Thesis, University of Helsinki, pp. 1–16.

Koskimies, O. & Saxén, L. (1966). Studies on kidney tubulogenesis. IV. Lactic dehydrogenase isozymes in the development of mouse metanephrogenic mesenchyme *in vitro*. *Ann. Med. exp. Fenn.* **44**, 151–4.

Kotani, M. (1962). The differentiation of Wolffian duct, mesonephros, and primordial germ cells in the newt *Triturus pyrrhogaster* after extirpation of the primordium of Wolffian duct. *J. Biol. Osaka City Univ.* **13**, 111–18.

Kratochwil, K. (1972). Tissue interaction during embryonic development. In *Tissue Interactions and Carcinogenesis*, ed. D. Tarin, pp. 1–47. Academic Press, London.

Kratochwil, K. (1977). Development and loss of androgen responsiveness in the embryonic rudiment of the mouse mammary glands. *Devel. Biol.* **61**, 358–65.

Kratochwil, K. (1983). Embryonic induction. In *Cell Interactions and Development. Molecular Mechanisms*, ed. K. M. Yamada, pp. 100–22. John Wiley & Sons, New York.

Kratochwil, K., Durnberger, H., Heuberger, B. & Wasner, G. (1979). Hormone-induced cell and tissue interaction in the embryonic mammary gland. *Cold Spring Harbor Conference on Cell Proliferation* **6**, 717–26.

Kreisberg, J. I. & Karnovsky, M. J. (1983). Glomerular cells in culture. *Kidney Int.* **23**, 439–47.

Kurtz, S. M. (1958). The electron microscopy of the developing human renal glomerulus. *Exp. Cell Res.* **14**, 355–67.

Lahti, A. & Saxén, L. (1966). Studies on kidney tubulogenesis. VIII. Appearance of kidney-specific antigens during *in vivo* and *in vitro* development of secretory tubules. *Exp. Cell Res.* **44**, 563–71.

Laitinen, L., Virtanen, I. & Saxén, L. (1986). Changes in the glycosylation pattern during embryonic development of mouse kidney as revealed with lectin-conjugates. *J. Histochem. Cytochem.* **34**, (in press).

Landschulz, W., Thesleff, I. & Ekblom, P. (1984). A lipophilic iron chelator can replace transferrin as a stimulator of cell proliferation and differentiation. *J. Cell Biol.* **98**, 596–601.

Lash, J. W. (1963). Studies on the ability of embryonic mesonephros explants to form cartilage. *Devel. Biol.* **6**, 219–32.

Lash, J. W. & Saxén, L. (1972). Human teratogenesis: *in vitro* studies on thalidomide inhibited chondrogenesis. *Devel. Biol.* **28**, 61–70.

Lash, J. W. & Vasan, N. (1978). Somite chondrogenesis *in vitro*. Stimulation by exogenous matrix components. *Devel. Biol.* **66**, 151–71.

Lash,. J., Saxén, L. & Ekblom, P. (1983). Biosynthesis of proteoglycans in organ cultures of developing kidney mesenchyme. *Exp. Cell Res.* **147**, 85–93.

Lawson, K. A. (1972). The role of mesenchyme in the morphogenesis and functional differentiation of rat salivary epithelium. *J. Embryol. Exp. Morphol.* **27**, 497–513.

Lawson, K. A. (1974). Mesenchyme specificity in rodent salivary gland development: the response of salivary epithelium to lung mesenchyme *in vitro*. *J. Embryol. Exp. Morphol.* **32**, 469–93.

Lazarides, E. (1982). Intermediate filaments: a chemically heterogeneous, developmentally regulated class of proteins. *Annu. Rev. Biochem.* **51**, 219–50.

Le Douarin, N. (1973). A biological cell labelling technique and its use in experimental embryology. *Devel. Biol.* **30**, 217–22.

Le Douarin, N. & Barq, G. (1969). Sur l'utilisation des cellules de la caille japonaise comme 'marqueurs biologiques' en embryologie expérimentale. *C. R. Acad. Sci. Paris* **269**, 1443–6.

Le Douarin, N. & Teillet, M.-A. M. (1974). Experimental analysis of the migration and differentiation of neuroblasts of the autonomic nervous system and of neurectodermal mesenchymal derivatives, using a biological cell marking technique. *Devel. Biol.* **41**, 162–84.

Leeson, T. S. (1971). An electron microscopic study of the post-natal development of the hamster kidney. With particular reference to the intertubular tissue. *Lab. Invest.* **10**, 466–80.

Lehtonen, E. (1975). Epithelio-mesenchymal interface during mouse kidney tubule induction *in vivo*. *J. Embryol. Exp. Morphol.* **34**, 695–705.

Lehtonen, E. (1976). Transmission of signals in embryonic kidney. *Med. Biol.* **54**, 108–28.

Lehtonen, E. & Saxén, L. (1986*a*). Control of differentiation. In *Human Growth*, ed. F. Falkner & J. M. Tanner, pp. 27–51. Plenum Press, New York.

Lehtonen, E. & Saxén, L. (1986*b*). Cytodifferentiation vs. organogenesis in kidney development. In *Progress in Developmental Biology*, ed. H. Slavkin, pp. 411–18. Alan R. Liss, New York.

Lehtonen, E., Wartiovaara, J., Nordling, S. & Saxén, L. (1975). Demonstration of cytoplasmic processes in Millipore filters permitting kidney tubule induction. *J. Embryol. Exp. Morphol.* **33**, 187–203.

Lehtonen, E., Jalanko, H., Laitinen, L., Miettinen, A., Ekblom, P. & Saxén, L. (1983). Differentiation of metanephric tubules following a short induction pulse. *Roux' Arch.* **192**, 145–51.
Lehtonen, E., Virtanen, I. & Saxén, L. (1985). Reorganization of intermediate cytoskeleton in induced metanephric mesenchyme cells is independent of tubule morphogenesis. *Devel. Biol.* **108**, 481–90.
Leivo, I., Vaheri, A., Timpl, R. & Wartiovaara, J. (1980). Appearance and distribution of collagens and laminin in the early mouse embryo. *Devel. Biol.* **76**, 100–14.
Lewis, O. J. (1958). The development of the blood vessels of the metanephros. *J. Anat.* 92, 84–97.
Linder, E. (1969). Differentiation of kidney antigen in human foetus. *J. Embryol. Exp. Morphol.* **21**, 517–37.
Linsenmayer, T. F., Trelstad, R. L. & Gross, J. (1973). The collagen of chick embryonic notochord. *Biochem. Biophys. Res. Commun.* **53**, 39–45.
Lombard, M.-N. & Grobstein, C. (1969). Activity in various embryonic and postembryonic sources for induction of kidney tubules. *Devel. Biol.* **19**, 41–51.
Ludwig, E. (1962). Uber Frühstadien der menschlichen Ureterbaumes. *Acta Anat.* **49**, 168–85.
Machemer, H. (1929). Differenzierungsfähigkeit der Urinierenanlage von *Triton alpestris. Roux' Arch.* **118**, 200–51.
Madri, J. A., Roll, F. J., Furthmayer, H. & Foidart, J. M. (1980). Ultrastructural localization of fibronectin and laminin in the basement membranes of murine kidney. *J. Cell Biol.* **86**, 682–7.
Magre, S. & Jost, A. (1984). Dissociation between testicular organogenesis and endocrine cytodifferentiation of Sertoli cells. *Proc. Nat. Acad. Sci. USA* **81**, 7831–4.
Marin-Padilla, M. (1964). The mesonephric–testicular connection in man and some mammals. *Anat. Rec.* **148**, 1–14.
Markert, C. L. & Ursprung, H. (1962). The ontogeny of isozyme patterns of lactate dehydrogenase in the mouse. *Devel.Biol.* **5**, 363–81.
Martin, C. (1976). Etude chez les oiseaux de l'influence du mésenchyme néphrogène sur le canal de Wolff à l'aide d'associations hétérospécifiques. *J. Embryol. Exp. Morphol.* **35**, 485–98.
Martin, C., Croisille, Y. & Gumpel-Pinot, M. (1971). Sur l'organogenèse du mésonéphros chez les oiseaux: potentialités évolutives du mésenchyme mésonéphrogène de l'embryon de poulet. *C.R. Acad. Sci. Paris* **272**, 863–4.
Maschkowzeff, A. (1936). Entfernung, Transplantation und Entwicklung der Keimanlage des Pronephros *in vitro* bei *Siredon pisciformis* und *Rana temporaria. Zool. Jb. Allg. Zool. Phys.* **54**, 1–40.
Miettinen, A. & Linder, A. (1976). Membrane antigens shared by renal proximal tubules and other epithelia associated with absorption and excretion. *Clin. Exp. Immunol.* **23**, 568–77.
Miettinen, H., Ellem, K. A. O. & Saxén, L. (1966). Studies on kidney tubulogenesis. VII. The response of RNA synthesis of mouse metanephrogenic mesenchyme to an inductive stimulus. *Ann. Med. exp. Fenn.* **44**, 109–16.
Minuth, W. W. (1982). Cell associated glycoproteins synthesized by cultured renal tubular cells. *Histochemistry* **76**, 89–106.
Minuth, W. W. (1983). Induction and inhibition of outgrowth and development of renal collecting duct epithelium. *Lab. Invest.* **48**, 543–8.
Minuth, W. W. & Kriz, W. (1982). Culturing of renal collecting duct epithelium as globular bodies. *Cell Tissue Res.* **224**, 335–48.
Minuth, W. W., Lauer, G. & Kriz, W. (1984). Immunocytochemical localization of a

renal glycoprotein ($gp_{CD}I$) synthesized by cultured collecting duct cells. *Histochemistry* **80**, 171–82.
Mittenthal, J. E. & Mazo, R. M. (1983). A model for shape generation by strain and cell-cell adhesion in the epithelium of an arthropod leg segment. *J. Theor. Biol.* **100**, 443–83.
Miura, K. (1930). Uber die einflüsse der totalen Extirpation des äusseren Glomerulus auf die Vorniere bei Froschlarven. *Jap. J. Med. Sci. Anat.* **2**, 125–33.
Moscona, A. A. (1974). Surface recognition of embryonic cells: lectin receptors, cell recognition and specific cell ligands. In *The Cell Surface in Development*, ed. A. A. Moscona, pp. 67–99. John Wiley & Sons, New York.
Neiss, W. F. (1982). Morphogenesis and histogenesis of the connecting tubule in the rat kidney. *Anat. Embryol.* **165**, 81–95.
Nieuwkoop, P. D. (1947). Experimental investigations on the origin and determination of the germ cells, and on the development of the lateral plates and germ ridges in urodeles. *Arch. Néerl. Zool.* **8**, 1–205.
Nordling, S., Miettinen, H., Wartiovaara, J. & Saxén, L. (1971). Transmission and spread of embryonic induction. I. Temporal relationship in transfilter induction of kidney tubules *in vitro*. *J. Embryol. Exp. Morphol.* **26**, 231–52.
Nordling, S., Ekblom, P., Lehtonen, E., Wartiovaara, J. & Saxén, L. (1978). Metabolic inhibitors and kidney tubule induction. *Med. Biol.* **56**, 372–9.
Nörgaard, J. O. R. (1983). Cellular outgrowth from isolated glomeruli. Origin and characterization. *Lab. Invest.* **48**, 526–42.
Oberley, T. D. & Steinert, B. W. (1983). Effect of the extracellular matrix molecules fibronectin and laminin on the adhesion and growth of primary renal cortical epithelial cells. *Virchows Arch. (Cell Pathol.)* **44**, 337–54.
Oberley, T. D., Mosher, D. F. & Mills, M. D. (1979). Localization of fibronectin within renal glomerulus and its production by cultured glomerular cells. *Am. J. Pathol.* **96**, 651–62.
O'Connor, R. J. (1938). Experiments on the development of the pronephric duct. *J. Anat.* **73**, 145–54.
O'Connor, R. J. (1939). Experiments on the development of the amphibian mesonephros. *J. Anat.* **74**, 34–44.
Okada, T. S. (1965). Changes in antigenic constitutions of embryonic chicken kidney cells during *in vitro* spreading culture. *Exp. Cell Res.* **39**, 591–603.
Okada, T. S. & Sato, A. G. (1963). Soluble antigens in microsomes of adult and embryonic kidneys. *Exp. Cell Res.* **31**, 251–65.
Oliver, J. (1939). *Architecture of the Kidney in Bright's Chronic Disease*. Paul B. Hoeber, Inc., New York.
Osathanondh, V. & Potter, E. (1963*a*). Development of human kidney as shown by microdissection. I. Preparation of tissue with reasons for possible misinterpretations of observations. *Arch. Pathol.* **76**, 271–6.
Osathanondh, V. & Potter, E. (1963*b*). Development of human kidney as shown by microdissection. II. Renal pelvis, calyces, and papillae. *Arch. Pathol.* **76**, 277–89.
Osathanondh, V. & Potter, E. (1963*c*). Development of human kidney as shown by microdissection. III. Formation and interrelationships of collecting tubules and nephrons. *Arch. Pathol.* **76**, 290–302.
Osathanondh, V. & Potter, E. (1966). Development of human kidney as shown by microdissection. V. Development of vascular pattern of glomerulus. *Arch. Pathol.* **82**, 403–11.
Osman, A. & Ruch, J. V. (1978). Contribution à l'étude des paramètres du cycle cellulaire au cours de l'odontogènese chez la souris. *J. Biol. Buccale* **6**, 43–54.
Overton, J. (1959). Studies on the mode of outgrowth of the amphibian pronephric duct. *J. Embryol. Exp. Morphol.* **7**, 86–93.

Paatela, M (1963). Renal microdissection in infants. With special reference to the congenital nephrotic syndrome. Ph.D. thesis. Mercatorin Kirjapaino, Helsinki.
Pasteels, J. (1942). New observations concerning the maps of presumptive areas of the young amphibian gastrula (*Amblystoma* and *Discoglossus*). *J. Exp. Zool.* **89**, 255–81.
Patsavoudi, E., Magre, S., Castanier, M., Scholler, R. & Jost, A. (1985). Dissociation between testicular morphogenesis and functional differentiation of Leydig cells. *J. Endocr.* **105**, 235–8.
Perantoni, A., Kan., F.W.-K., Dove, F. F. & Reed, C. D. (1985). Selective growth in culture of fetal rat renal collecting duct anlagen: morphologic and biochemical characterization. *Lab. Invest.* **53**, 589–96.
Peter, K. (1927). *Untersuchungen über Bau und Entwicklung der Niere.* Gustav Fischer, Jena.
Peterson, C. A. & Grainger, R. M. (1985). Differentiation of embryonic chick feather-forming and scale-forming tissues in transfilter cultures. *Devel. Biol.* **111**, 8–25.
Pictet, R. L., Filosa, S., Phelps, P. & Rutter, W. J. (1975). Control of DNA synthesis in the embryonic pancreas: interaction of the mesenchymal factor and cyclic AMP. In *Extracellular Matrix Influences on Gene Expression*, ed. H. C. Slavkin & R. C. Greulich, pp. 531–40. Academic Press, New York.
Poole, T. J. & Steinberg, M. S. (1981). Amphibian pronephric duct morphogenesis: segregation, cell rearrangement and directed migration of the *Ambystoma* duct rudiment. *J. Embryol. Exp. Morphol.* **63**, 1–16.
Poole, T. J. & Steinberg, M. S. (1982). Evidence for the guidance of pronephric duct migration by a craniocaudally traveling adhesion gradient. *Devel. Biol.* **92**, 144–58.
Potter, E. L. (1965). Development of the human glomerulus. *Arch. Path.* **80**, 241–55.
Preminger, G. M., Koch, W. F., Fried, F. A. & Mandell, J. (1980). Utilization of the chick chorioallantoic membrane for *in vitro* growth of the embryonic murine kidney. *Am. J. Anat.* **159**, 17–24.
Preminger, G. M., Koch, W. E., Fried, F. A. & Mandell, J. (1981). Chorioallantoic membrane grafting of the embryonic murine kidney. An improved *in vitro* technique for studying kidney morphogenesis. *Investig. Urol.* **18**, 377–81.
Price, G. C. (1897). Development of the excretory organs of a myxinoid, *Bdellostoma stouti*, Lockington. *Zool. Jb. Anat. Ontog.* **10**, 205–26.
Rapola, J. & Niemi, M. (1965). Studies on kidney tubulogenesis. Cytochemical localization of phosphatase and dehydrogenase activities during the formation of tubules *in vitro. Z. Anat. Entwicklungsgesch.* **124**, 309–20.
Rappaport, R. (1955). The initiation of pronephric function in *Rana pipiens. J. Exp. Zool.* **128**, 481–8.
Reeves, W. H., Kanwar, Y. P. & Farquhar, M. G. (1980). Assembly of the glomerular filtration surface. Differentiation of anionic sites in glomerular capillaries of newborn rat kidney. *J. Cell Biol.* **85**, 735–53.
Remak, R. (1855). *Untersuchungen über die Entwicklung der Wirbeltiere.* Reimer, Berlin.
Richter, A., Sanford, K. K. & Evans, V. J. (1972). Influence of oxygen and culture media on plating efficiency of some mammalian tissue cells. *J. Natl. Cancer Inst.* **49**, 1705–12.
Rienhoff, W. F. (1922). Development and growth of the metanephros or permanent kidney in chick embryos. *Johns Hopkins Hosp. Bull.* **33**, 392–406.
Risau, W. & Ekblom, P. (1986). Production of a heparin-binding angiogenesis factor by the embryonic kidney. *J. Cell Biol.* **103**, 1101–7.
Ronzio, R. A. & Rutter, W. J. (1973). Effects of partially purified factor from chick embryos on macromolecular synthesis of embryonic pancreatic epithelia. *Devel. Biol.* **30**, 307–20.

Runner, M. N. (1946). The development of the mesonephros of the albino rat in intraocular grafts. *J. Exp. Zool.* **103**, 305–19.
Ruoslahti, E., Engvall, E. & Hayman, E. (1981). Fibronectin: current concepts of its structure and functions. *Coll. Rel. Res.* **1**, 85–128.
Rutenberg, A. M., Kim, H., Fischbein, J. W., Hanker, J. S., Wasserkrug, H. L. & Seligman, A. M. (1969). Histochemical and ultrastructural demonstration of γ-glutamyl transpeptidase activity. *J. Histochem. Cytochem.* **17**, 517–26.
Rutter, W. J., Wessells, N. K. & Grobstein, C. (1964). Control of specific synthesis in the developing pancreas. *Nat. Cancer Inst. Monogr.* **13**, 51–65.
Rutter, W. J., Ball, W. D., Bradshaw, W. S., Clark, W. R. & Sanders, T. G. (1967). Levels of regulation in cytodifferentiation. In *Morphological and Biochemical Aspects of Cytodifferentiation,* ed. E. Hagen, W. Wechsler & F. Zilliken, pp. 110–24. Karger, Basel.
Salama, J., Folio, P. & Chevrel, J. P. (1982). Reconstruction du métanéphros chez un embryon humain de 20 millimètres (VC) étude de l'origine de l'artère rénale. *Bull. Assoc. Anat.* **66**, 397–406.
Salzgeber, B. & Weber, R. (1966). La régression du mésonéphros chez l'embryon de poulet. Etude des activités de la phosphatase acide et des cathépsines. Analyse biochimique, histochimique et observations au microscope électronique. *J. Embryol. Exp. Morphol.* **15**, 397–419.
Sariola, H. (1984). Incomplete fusion of the epithelial and endothelial basement membranes in interspecies hybrid glomeruli. *Cell Different.* **14**, 189–95.
Sariola, H. (1985). Interspecies chimeras: an experimental approach for studies on the embryonic angiogenesis. *Med. Biol.* **63**, 43–65.
Sariola, H., Ekblom, P., & Saxén, L. (1982). Restricted developmental options of the metanephric mesenchyme. In *Embryonic Development,* part B: *Cellular Aspects*, ed. M. Burger & R. Weber, pp. 425–31. Alan R. Liss, Inc., New York.
Sariola, H., Ekblom, P., Lehtonen, E. & Saxén, L. (1983). Differentiation and vascularization of the metanephric kidney grafted on the chorioallantoic membrane. *Devel. Biol.* **96**, 427–35.
Sariola, H. Kuusela, P. & Ekblom, P. (1984*a*). Cellular origin of fibronectin in interspecies hybrid kidneys. *J. Cell Biol.* **99**, 2099–107.
Sariola, H., Timpl, R., von der Mark, K., Mayne, R., Fitch, J. M., Linsenmayer, T. F. & Ekblom, P. (1984*b*). Dual origin of glomerular basement membrane. *Devel. Biol.* **101**, 86–96.
Sariola, H., Peault, B., Le Douarin, N., Buck, C., Dieterlen, F. & Saxén, L. (1984*c*). Extracellular matrix and capillary ingrowth in interspecies chimeric kidneys. *Cell Different.* **15**, 43–52.
Saxén, L. (1961). Transfilter neural induction of amphibian ectoderm. *Devel. Biol.* **3**, 140–52.
Saxén, L. (1970*a*). Failure to demonstrate tubule induction in a heterologous mesenchyme. *Devel. Biol.* **23**, 511–23.
Saxén, L. (1970*b*). The determination and differentiation of the metanephric nephron. In *Embryology, Ultrastructure Physiology,* Proc. Fourth Int. Congr. Nephrol, vol. 1, ed. N. Alwall, F. Berglund & B. Josephson, pp. 29–38. S. Karger, Basel.
Saxén, L. (1971). Inductive interactions in kidney development. In *Control Mechanisms of Growth and Differentiation*, XXV Symposia of the Society for Experimental Biology, ed. D. D. Davies & M. Balls, pp. 207–21. Cambridge University Press, Cambridge.
Saxén, L. (1972). Interactive mechanisms in morphogenesis. In *Tissue Interactions and Carcinogenesis*, ed. D. Tarin, pp. 49–80. Academic Press, London.
Saxén, L. (1977). Directive versus permissive induction: a working hypothesis. In *Cell*

and Tissue Interactions, ed. J. W. Lash & M. M. Burger, pp. 1–9. Raven Press, New York.

Saxén, L. (1980). Mechanism of morphogenetic tissue interactions: the message of transfilter experiments. In *Results and Problems in Cell Differentiation*, vol. 11: *Differentiation and Neoplasia*, ed. R. G. McKinnell, M. A. DiBerardino, M. Blumenfeld & R. D. Bergad, pp. 147–54. Springer-Verlag, Berlin & Heidelberg.

Saxén, L. (1981). Induction in embryogenesis. *Verh. Dtsch. Zool. Ges.* 146–9.

Saxén, L. (1983). *In vitro* model-systems for chemical teratogenesis. In *In Vitro Toxicity Testing of Environmental Agents*, part B, ed. A. Kolber, T. K. Wong, L. D. Grant, R. S. DeWoskin & T. J. Hughes, pp. 173–90. Plenum Press, New York.

Saxén, L. (1984*a*). Implementation of a developmental program. In *Modern Biological Experimentation*, ed. C. Chagas, pp. 155–63. Pontificia Academia Scientiarum, Città del Vaticano.

Saxén, L. (1984*b*). Chimeric tissue combinations in the analysis of developmental mechanisms in the embryonic kidney. In *Chimeras in Developmental Biology*, ed. N. Le Douarin & A. McLaren, pp. 401–8. Academic Press, London.

Saxén, L. & Karkinen-Jääskeläinen, M. (1975). Inductive interactions in morphogenesis. In *The Early Development of Mammals*, ed. M. Balls & A. Wild, pp. 319–34. Cambridge University Press, Cambridge.

Saxén, L. & Kohonen, J. (1969). Inductive tissue interactions in vertebrate morphogenesis. *Int. Rev. Exp. Pathol.* **8**, 57–128.

Saxén, L. & Lehtonen, E. (1978). Transfilter induction of kidney tubules as a function of the extent and duration of intercellular contacts. *J. Embryol. Exp. Morphol.* **47**, 97–109.

Saxén, L. & Lehtonen, E. (1986). Cells into organs. In *Co-ordinated Regulation of Gene Expression*, ed. R. Clayton & D. E. S. Truman, pp. 269–78. Plenum Press, New York.

Saxén, L. & Saksela, E. (1971). Transmission and spread of embryonic induction. II. Exclusion of an assimilatory transmission mechanism in kidney tubule induction. *Exp. Cell Res.* **66**, 369–77.

Saxén, L. & Toivonen, S. (1961). The two-gradient hypothesis in primary induction. The combined effect of two types of inductors mixed in different ratios. *J. Embryol. Exp. Morphol.* **9**, 514–33.

Saxén, L. & Toivonen, S. (1962). *Primary Embryonic Induction*. Academic Press, London.

Saxén, L. & Wartiovaara, J. (1966). Cell contact and cell adhesion during tissue organization. *Int. J. Cancer* **1**, 271–90.

Saxén, L., Vainio, T. & Toivonen, S. (1962). Effect of polyoma virus on mouse kidney rudiment *in vitro*. *J. Nat. Cancer Inst.* **29**, 597–631.

Saxén, L., Toivonen, S., Vainio, T. & Korhonen, P. (1965*a*). Untersuchungen über die Tubulogenese der Niere. III. Die Analyse der Frühentwicklung mit der Zeitraffermethode. *Z. Naturforsch.* **20b**, 340–3.

Saxén, L., Wartiovaara, J., Häyry, P. & Vainio, T. (1965*b*). Cell contact and tissue interaction in cytodifferentiation. In *Rep. IV Scand. Congr. Cell Res.*, ed. A. Brögger, pp. 21–36. Universitetsförlaget, Oslo.

Saxén, L., Koskimies, O., Lahti, A., Miettinen, H., Rapola, J. & Wartiovaara, J. (1968). Differentiation of kidney mesenchyme in an experimental model system. In *Advances in Morphogenesis*, vol. 7, ed. M. Abercrombie, J. Brachet & T. J. King, pp. 251–93. Academic Press, London.

Saxén, L., Karkinen-Jääskeläinen, M., Lehtonen, E., Nordling, S. & Wartiovaara, J. (1976*a*). Inductive tissue interactions. In *The Cell Surface in Animal Embryogenesis and Development*, ed. G. Poste & G. L. Nicolson, pp. 331–407. North-Holland Publ. Co., Amsterdam.

Saxén, L., Lehtonen, E., Karkinen-Jääskeläinen, M., Nordling, S. & Wartiovaara, J. (1976*b*). Are morphogenetic tissue interactions mediated by transmissible signal substances or through cell contacts? *Nature (Lond.)* **259**, 662–3.

Saxén, L., Ekblom, P. & Thesleff, I. (1980). Mechanism of morphogenetic cell interactions. In *Development in Mammals*, vol. 4, ed. M. H. Johnson, pp. 161–202. Biomedical Press, Elsevier/North Holland Publ. Co., Amsterdam.

Saxén, L., Ekblom, P. & Lehtonen, E. (1981). The kidney as a model system for determination and differentiation. In *Biology of Normal Human Growth*, ed. M. Ritzen, K. Hall, A. Zetterberg, A. Aperia, A. Larsson & R. Zetterström, pp. 117–27. Raven Press, New York.

Saxén, L., Ekblom, P. & Thesleff, I. (1982). Cell matrix interactions in organogenesis. In *New Trends in Basement Membrane Research*, ed. K. Kuehn, H. Schoene & R. Timpl, pp. 259–66. Raven Press, New York.

Saxén, L., Salonen, J., Ekblom, P. & Nordling, S. (1983). DNA synthesis and cell generation cycle during determination and differentiation of the metanephric mesenchyme. *Devel. Biol.* **98**, 130–8.

Saxén, L., Ekblom, P. & Sariola, H. (1985). Organogenesis. In *Prevention of Physical and Mental Congenital Defects*, ed. M. Marois, pp. 41–53. Alan R. Liss, Inc., New York.

Schiller, A. & Tiedemann, K. (1981). The mature mesonephric nephron of the rabbit embryo. III. Freeze-fracture studies. *Cell Tissue Res.* **221**, 431–42.

Schor, A. & Schor, S. (1983). Tumour angiogenesis. *J. Pathol.* **141**, 385–413.

Seevers, C. H. (1932). Potencies of the end bud and other caudal levels of the early chick embryo, with special reference to the origin of the metanephros. *Anat. Rec.* **54**, 217–46.

Shimasaki, Y. (1930). Entwicklungsmechanische Untersuchungen über die Uriniere des Bufo. *Jap. J. Med. Sci. Anat.* **2**, 291–319.

Shin-Iké, T. (1955). Further experiments on the development of the pronephric duct in amphibia. *Annot. Zool. Jap.* **138**, 215–26.

Sikri, K. L., Foster, C. L., Bloomfield, F. L. & Marshall, R. D. (1979). Localization by immunofluorescence and light- and electronmicroscopic immunoperoxidase techniques of Tamm–Horsfall glycoprotein in adult hamster kidney. *Biochem. J.* **181**, 525–32.

Silverman, H. (1969). Uber die Entwicklung der Epithelplatten in den *Corpuscula renalia* der menschlichen Uriniere. *Acta Anat.* **74**, 36–43.

Sobel, J. S. (1966). DNA synthesis and differentiation in embryonic kidney mesenchyme *in vitro*. *Science* **153**, 1387–9.

Spemann, H. (1901). Ueber Korrelationen in der Entwicklung des Auges. *Verh. Anat. Ges. (Jena)* **15**, 61–79.

Spemann, H. (1912). Zur Entwicklung des Wirbeltierauges. *Zool. Jb. Allgem. Zool. Physiol. Tiere* **32**, 1–98.

Spemann, H. (1936). *Experimentelle Beiträge zu einer Theorie der Entwicklung*. Springer Verlag, Berlin.

Spofford, W. R. (1945). Observations on the posterior part of the neural plate in *Ambystoma. J. Exp. Zool.* **9**, 35–52.

Spofford, W. R. (1948). Observations on the posterior part of the neural plate in *Amblystoma*. II. The inductive effect of the intact part of the chordamesodermal axis on competent prospective ectoderm. *J. Exp. Zool.* **107**, 123–59.

Spooner, B. S. & Hilfer, S. R. (1971). The expression of differentiation by chick embryo thyroid in cell culture. II. Modication of phenotype in monolayer cultured by different media. *J. Cell Biol.* **48**, 225–34.

Spooner, B. S., Cohen, H. I. & Faubion, J. (1977). Development of the embryonic mammalian pancreas: the relationship between morphogenesis and cytodifferentiation. *Devel.. Biol.* **61**, 119–30.

Stampfli, H. R. (1950). Histologische Studien am Wolffischen Körper (Mesonephros) der Vögel und über seinen Umbau zu Nebenhoden und Nebenovar. *Rev. Suiss. Zool.* **57**, 237–315.

Stenman, S. & Vaheri, A. (1978). Distribution of a major connective tissue protein, fibronectin, in normal human tissues. *J. Exp. Med.* **147**, 1054–64.

Steward, P. A. & Wiley, M. J. (1981). Developing nervous tissue induces formation of blood–brain barrier characteristics in invading endothelial cells: a study using quail–chick transplantation chimeras. *Devel. Biol.* **84**, 183–92.

Striker, G. E., Killen, P. D. & Farin, F. M. (1980). Human glomerular cells *in vitro*: isolation and characterization. *Transplant. Proc.* Suppl. **1**, 88–99.

Strudel, G. & Pinot, M. (1965). Differenciation en culture *in vitro* du mesonephros de l'embryon de poulet. *Devel. Biol.* **1**, 284–99.

Sundelin, P., Wartiovaara, J., Saxén, L. & Thorell, B. (1969). Increased cytoplasmic RNA reflecting an early step of inductive tissue interaction. *Exp. Cell Res.* **54**, 347–52.

Suzuki, Y. (1959). An electron microscopy of renal differentiation. II. Glomerulus. *Keio J. Med.* **8**, 129–43.

Thesleff, I. & Ekblom, P. (1984). Role of transferrin in branching morphogenesis, growth and differentiation of the embryonic kidney. *J. Embryol. Exp. Morphol.* **82**, 147–61.

Thesleff, I., Lehtonen, E., Wartiovaara, J. & Saxén, L. (1977). Interference of tooth differentiation with interposed filters. *Devel. Biol.* **58**, 197–203.

Thesleff, I., Lehtonen, E. & Saxén, L. (1978). Basement membrane formation in transfilter tooth culture and its relation to odontoblast differentiation. *Differentiation* **10**, 71–9.

Thesleff, I., Stenman, S., Vaheri, A. & Timpl, R. (1979). Changes in the matrix proteins fibronectin and collagen of mouse tooth germ. *Devel. Biol.* **70**, 116–26.

Thesleff, I., Ekblom, P., Kuusela, P., Lehtonen, E. & Ruoslahti, E. (1983). Exogenous fibronectin is not required for organogenesis *in vitro*. *In Vitro* **19**, 903–10.

Thesleff, I., Partanen, A.-M., Landschulz, W., Trowbridge, I. S. & Ekblom, P. (1985). The role of transferrin receptors and iron delivery in mouse embryonic morphogenesis. *Differentiation* **30**, 152–8.

Thiery, J.-P. (1984). Mechanism of cell migration in the vertebrate embryo. *Cell Different.* **15**, 1–15.

Thiery, J.-P., Duband, J. L. & Delouvée, A. (1982*a*). Pathways and mechanism of avian trunk neural crest cell migration and localization. *Devel. Biol.* **93**, 324–43.

Thiery, J.-P., Duband, J. P., Rutishauser, U. & Edelman, G. M. (1982*b*). Cell adhesion molecules in early chicken embryogenesis. *Proc. Nat. Acad. Sci. USA* **79**, 6737–41.

Thiery, J.-P., Delouvée, A., Gallin, W. J., Cunningham, B. A. & Edelman, G. M. (1984). Ontogenetic expression of cell adhesion molecules: L-CAM is found in epithelia derived from the three primary germ layers. *Devel. Biol.* **102**, 61–78.

Thorning, D. & Vracko, R. (1977). Renal glomerular basal lamina scaffold: embryonic development, anatomy, and role in cellular reconstruction of rat glomeruli injured by freezing and thawing. *Lab. Invest.* **37**, 105–19.

Ti-Chow-Tung & Su-Hwei-Ku (1944). Experimental studies on the development of the pronephric duct in anuran embryos. *J. Anat.* **78**, 52–7.

Tiedemann, K. (1976). The mesonephros of cat and sheep. Comparative morphological and histochemical studies. In *Advances in Anatomy, Embryology and Cell Biology*, ed. A. Brodal, W. Hild, J. van Limborgh, R. Ortmann, T. H. Schiebler, G. Töndury & E. Wolff, pp. 1–119. Springer-Verlag, Berlin.

Tiedemann, K. (1979). Architecture of the mesonephric nephron in pig and rabbit. *Anat. Embryol.* **157**, 105–12.
Tiedemann, K. (1983). The pig mesonephros. I. Enzyme histochemical observations on the segmentation of the nephron. *Anat. Embryol.* **137**, 113–23.
Tiedemann, K. & Egerer, G. (1984). Vascularization and glomerular ultrastructure in the pig mesonephros. *Cell Tissue Res.* **238**, 165–75.
Tiedemann, K. & Wettstein, R. (1980). The mature mesonephric nephron of the rabbit embryo. I. SEM-studies. *Cell Tissue Res.* **209**, 95–109.
Tiedemann, K. & Zaar, K. (1983). The pig mesonephros. II. The proximal tubule: SEM, TEM and freeze-fracture images. *Anat. Embryol.* **168**, 241–52.
Timpl, R. & Martin, G. R. (1982). Components of basement membranes. In *Immunochemistry of the Extracellular Matrix*, vol. II: *Applications*, ed. H. Furthmayer, pp. 119–50. CRC Press Inc., Boca Raton, Florida.
Timpl, R., Rohde, H., Robey, P. G., Rennard, S. I., Foidart, J. M. & Martin, G. R. (1979). Laminin – a glycoprotein from basement membranes. *J. Biol. Chem.* **254**, 9933–7.
Timpl, R., Wiedemann, H., Van Delden, V., Furthmayer, H. & Kuehn, K. (1981). A network model for the organization of type IV collagen molecules in basement membranes. *Eur. J. Biochem.* **120**, 203–11.
Timpl, R., Dziadek,. M., Fujiwara, H., Nowak, H. & Wick, G. (1983). Nidogen, a new self-aggregating basement membrane protein. *Eur. J. Biochem.* **137**, 455–54.
Toivonen, S. (1945). Uber die Entwicklung der Vor- und Uriniere beim Kaninchen. *Ann. Acad. Sci. Fenn.* ser. A, **8**, 1–27.
Toivonen, S. (1979). Transmission problem in primary induction. *Differentiation* **15**, 177–81.
Toivonen, S. & Saxén, L. (1955). The simultaneous inducing action of liver and bone-marrow of the guinea pig in implantation and explantation experiments with embryos of *Triturus*. *Exp. Cell Res.* Suppl. **3**, 346–57.
Torrey, T. W. (1954). The early development of the human nephrons. *Contrib. Embryol. Carnegie Inst. Wash.* **35**, 175–97.
Torrey, T. W. (1965). Morphogenesis of the vertebrate kidney. In *Organogenesis*, ed. R. L. DeHaan & H. Ursprung, pp. 559–79. Holt, Rinehart and Winston, New York.
Trinkaus, J. P. (1984). *Cells into Organs. The Forces that Shape the Embryo.* Prentice-Hall Inc., Englewood Cliffs, NJ.
Turpen, J. B. & Knudson, C. M. (1982). Ontogeny of hematopoietic cells in *Rana pipiens*: precursor cell migration during embryogenesis. *Develop. Biol.* **89**, 138–51.
Unsworth, B. & Grobstein, C. (1970). Induction of kidney tubules in mouse metanephrogenic mesenchyme by various embryonic mesenchymal tissues. *Devel. Biol.* **21**, 547–56.
Vaheri, A. & Mosher, D. (1978). High molecular weight, cell surface glycoprotein (fibronectin) lost in malignant transformation. *Biochim. Biophys. Acta* **516**, 1–25.
Vainio, T., Jainchill, J., Clement, K. & Saxén, L. (1965). Studies on kidney tubulogenesis. VI. Survival and nucleic acid metabolism of differentiating mouse metanephrogenic mesenchyme *in vitro J. Cell Comp. Physiol.* **66**, 311–18.
van Deth, J. H. M. (1946). *Experimental Embryology in the Netherlands*. Elsevier Publ. Co., Amsterdam.
van Geertruyden, J. (1942). Quelques précisions sur le développement du pronéphros et de l'uretère primaire chez les amphibiens anoures. *Ann. Soc. Zool. Belg.* **73**, 180–95.
van Geertruyden, J. (1946). Recherches expérimentales sur la formation du mésonéphros chez les amphibiens anoures. *Arch. Biol.* **57**, 145–81.

Vernier, R. L. & Birch-Anderson, A. (1962). Studies on the human fetal kidney. I. Development of the glomerulus. *J. Pediat.* **60**, 754–68.
Vestweber, K., Kemler, R. & Ekblom, P. (1985). Cell-adhesion molecule uvomorulin during kidney development. *Devel. Biol.* **112**, 213–21.
von der Mark, K. & von der Mark, H. (1977). Immunological and biochemical studies of collagen type transition during *in vitro* chondrogenesis of chick limb mesodermal cells. *J. Cell Biol.* 73, 736–47.
Waddington, C. H. (1938). The morphogenetic function of a vestigial organ in the chick. *J. Exp. Biol.* **15**, 371–6.
Warren, B. A., Greenblatt, M. and Kommineni, V. R. C. (1972). Tumour angiogenesis: ultrastructure of endothelial cells in mitosis. *Brit. J. Exp. Pathol.* **53**, 215–24.
Wartiovaara, J. (1966*a*). Studies on kidney tubulogenesis. V. Electronmicroscopy of basement membrane formation *in vitro. Ann. Med. Exp. Fenn.* **44**, 140–50.
Wartiovaara, J. (1966*b*). Cell contacts in relation to cytodifferentiation in metanephrogenic mesenchyme *in vitro. Ann. Med. Exp. Fenn.* **44**, 469–503.
Wartiovaara, J., Lehtonen, E., Nordling, S. & Saxén, L. (1972). Do membrane filters prevent cell contacts? *Nature (Lond.)* **238**, 407–8.
Wartiovaara, J., Nordling, S., Lehtonen, E. & Saxén, L. (1974). Transfilter induction of kidney tubules: correlation with cytoplasmic penetration into Nuclepore filters. *J. Embryol. Exp. Morphol.* **31**, 667–82.
Wartiovaara, J., Stenman, S. & Vaheri, A. (1976). Changes in expression of fibroblast surface antigen (SFA) during cytodifferentiation and heterokaryon formation. *Differentiation* **5**, 85–9.
Wartiovaara, J., Leivo, I. & Vaheri, A. (1980). Matrix glycoproteins in early mouse development and in differentiation of teratocarcinoma cells. In *The Cell Surface: Mediator of Developmental Process*, ed. S. Subtelny & N. K. Wessells, pp. 305–24. Academic Press, New York.
Waterman, A. J. (1940). Growth and differentiation of kidney tissue of the rabbit embryo in omental grafts. *J. Morphol.* **67**, 369–85.
Weiss, L. & Nir, S. (1979). On the mechanisms of transfilter induction of kidney tubules. *J. Theor. Biol.* **78**, 11–20.
Weiss, P. (1947). The problem of specificity in growth and development. *Yale J. Biol. Med.* **19**, 235–78.
Wessells, N. K. (1963). Effects of extra-epithelial factors on the incorporation of thymidine by embryonic epidermis. *Exp. Cell Res.* **30**, 36–55.
Wessells, N. K. (1970). Mammalian lung development: interactions in formation and morphogenesis of tracheal buds. *J. Exp. Zool.* **175**, 455–66.
Wessells, N. K. (1977). *Tissue Interactions and Development*. W. A. Benjamin, Inc., Menlo Park.
Wharton, L. R. (1949). Double ureters and associated renal anomalies in early human embryos. *Contrib. Embryol. Carn. Inst. Wash.* **33**, 103–12.
Willier, B. H. (1924). The endocrine glands and the development of the chick. I. The effects of thyroid grafts. *Am. J. Anat.* **33**, 67–103.
Winick, M. & McCrory, W. W. (1968). Renal differentiation. A model for the study of development. *Birth Defects* **4**, 1–14.
Wolff, E. (1966). General introduction. General factors of embryonic differentiation. In *Cell Differentiation and Morphogenesis*, pp. 1–23. North-Holland Publ. Co., Amsterdam.
Wolff, E. (1969). Mécanismes inducteurs dans l'organogenèse du rein. In *Les Interactions Tissulaires au Cours de l'Organogenèse*, ed. E. Wolff, pp. 49–66. Dunod, Paris.
Wolff, E., Wolff, E. & Bishop-Calame, S. (1969). Explants of embryonic kidney:

techniques and applications. In *The Kidney*, ed. C. Rouiller & A. F. Muller, vol. II, pp. 1–82. Academic Press, New York & London.

Yoshida-Noro, C., Suzuki, N. & Takeichi, M. (1984). Molecular nature of the calcium-dependent cell–cell adhesion system in mouse teratocarcinoma and embryonic cells studied with a monoclonal antibody. *Devel. Biol.* **101**, 19–27.

Author index

Subject index